Beall
Rotational Molding

Glenn L. Beall

Rotational Molding

Design, Materials, Tooling, and Processing

Hanser Publishers, Munich

Hanser/Gardner Publications, Inc., Cincinnati

The Author:
Glenn L. Beall, Glenn Beall Plastics, Ltd., 32981 North River Road, Libertyville, IL 60048, USA

Distributed in the USA and in Canada by
Hanser/Gardner Publications, Inc.
6915 Valley Avenue, Cincinnati, Ohio 45244-3029, USA
Fax: (513) 527-8950
Phone: (513) 527-8977 or 1-800-950-8977
Internet: http:/7www.hansergardner.com
Distributed in all other countries by
Carl Hanser Verlag
Postfach 86 04 20, 81631 München, Germany
Fax: 49 (89) 98 12 64

The use of general descriptive names, trademarks, etc., in this publication, even if the former are not especially identified, is not to be taken as a sign that such names, as understood by the Trade Marks and Merchandise Marks Act, may accordingly be used freely by anyone.

While the advice and information in this book are believed to be true and accurate at the date of going to press, neither the authors nor the editors nor the publisher can accept any legal responsibility for any errors or omissions that may be made. The publisher makes no warranty, express or implied, with respect to the material contained herein.

Library of Congress Cataloging-in-Publication Data
Beall, Glenn.
Rotational molding : design, materials, tooling, and processing/Glenn Beall.
    p.   cm.
Includes bibliographical references and index.
ISBN 1-56990-260-7
1. Plastics—Molding.   I. Title.
TP1150.B388   1998
668.4'12—dc21   98-33646

Die Deutsche Bibliothek—CIP-Einheitsaufnahme
Beall, Glenn:
Rotational molding: design, materials, tooling, and processing/Beall.—Munich:
Hanser; Cincinnati: Hanser/Gardner, 1998
ISBN 3-446-18790-1

© Carl Hanser Verlag, Munich 1998
Typeset in the U.K. by Techset Composition Ltd., Salisbury
Printed and bound in Germany by Druckhaus "Thomas Müntzer", Bad Langensalza

*This book is dedicated to the memory of
Helen Giles and George R. Ryan
Inspiring Mentors*

Foreword

There are many ways to process and produce plastic parts. There are at least five or six major processes that will make hollow parts. Generally, these processes are not in competition. When the complete specifications and marketplace requirements of a given plastic part are considered, one process usually recommends itself clearly over all the others.

It is the purpose of the Association of Rotational Molders, and this book by our Technical Director, Glenn Beall, to inform the general plastics community that rotational molding is a major production process. Furthermore, when a given part indicates the use of rotational molding, its strengths and advantages usually preclude all other processes.

Many recent advances have made the process more flexible, dependable, predictable, and marketable. We hope this book increases your knowledge of, familiarity with, and trust in this very unique process.

Thomas M. Niland
President
Association of Rotational Molders

Preface

Welcome to the dynamic field of rotational molding. You are about to explore a unique plastics manufacturing process that has never before been so thoroughly explained in such easy-to-grasp terms.

A significant number of engineers pass through universities without receiving instruction in plastic materials and processing technology. This is a shocking situation considering the wide use of plastics in virtually all types of manufactured products. Most universities that provide plastics curriculums tend to concentrate on only such major segments of the industry as plastic materials technology, and processes such as injection and blow molding and extrusion and thermoforming. This is an understandable situation, because these large industries provide the majority of the jobs that universities train their students to fill. There are, however, many other less frequently used plastics processes that are equally important to the success of the manufacturing industry. Many of these newer processes are enjoying greater growth and more career opportunities than the older and better established segments of the plastics industries.

Rotational molding is a classic example of this situation. The modern rotational molding process has been in existence since the 1940s, but its advantages and capabilities have only recently been publicized. For understandable reasons, the technology of rotational molding has been developed by and remains concentrated among the molders, plastic materials manufacturers, mold makers, and machinery builders within the industry.

This technology has only recently found its way into trade literature, seminars, and universities. As a result, manufacturers of durable products are uncertain when to specify rotational molding, or how to use this process to its best advantage. This book has been written as a guide for academia, original equipment manufacturers, and those just joining the industry. No attempt is made to teach the reader how to perform rotational molding. The whole purpose of this book is to help the reader understand and use this manufacturing technique to its maximum capabilities.

The emphasis is on when to specify rotational molding and how to design and develop hollow plastic products that can be efficiently produced. The origins of the process, its present status, and its future prospects are reviewed. The different types of available plastic materials, molds, and processing machines are described in sufficient detail to allow the user to intelligently select the ideal combination for a new product. The factors affecting cost and quality are explained. Many of the details to be considered in selecting the ideal suppliers are reviewed.

Every effort has been made to simplify the reader's introduction to and understanding of this process. No attempt has been made to exhaustively explore the depths of plastic material or rotational molding technology. Where possible the use of trade terms has been avoided. The glossary of terms will be helpful in clarifying the meaning of unfamiliar terms encountered in this book, or while working with the industry.

The reader will gain a broader understanding of this process and the standards and practices of the industry, and will be in a better position to design and develop low-cost, high-quality, hollow plastic parts that work correctly on the first try.

Glenn L. Beall

Contents

1 The Rotational Molding Industry

1.1 Process Definition

Rotational molding is a high-temperature, low-pressure, open-molding plastic-forming process that uses heat and biaxial rotation to produce hollow, one-piece parts.

1.2 Process Description

To rotationally mold a roll-out refuse container, a mold that defines the shape of the part to be produced is mounted on the arm of a molding machine (Fig. 1.1). The machine is capable of biaxially rotating and moving the mold through the four phases of the process.

A predetermined amount of plastic material, in the form of a liquid or a powder, is then placed in the mold's cavity (Fig. 1.1A).

The machine then simultaneously rotates the mold in two directions and moves the mold into the heating chamber or oven (Fig. 1.1B). In the oven, the mold becomes hot and all the plastic material adheres to and sinters onto the inside surface of the cavity.

While it continues to rotate, the machine moves the mold out of the heating chamber and into the cooling chamber, where the plastic is cooled to the point that the formed plastic part will retain its shape (Fig. 1.1C).

The machine then moves the mold to the open station, and the mold stops rotating. The mold can then be opened and the molded part removed (Fig. 1.1D).

The mold is then recharged with plastic material and the process can be repeated.

There are many different types of single- and multiple-arm rotational molding machines now in use. The multi-arm turret machine (Fig. 1.2) is the most common type of machine in use today.

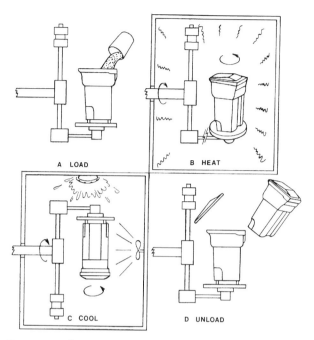

Figure 1.1 The four steps of the rotational molding process. A) Load plastic into mold. B) Heat mold and plastic. C) Cool mold and plastic. D) Remove the molded part

1.3   Attributes of the Process

Rotational molding has always been thought of as the ideal process for producing large plastic tanks. This process is, however, capable of producing many other types of parts (Fig. 1.3). Like all other manufacturing processes, rotational molding has its share of advantages and disadvantages.

1.3.1   Advantages

The advantages that drive the rotational molding process are its ability to produce large and small seamless, one-piece, hollow parts of extremely complex shapes. Many of these complex shapes, such as the sludge-lift tank with its curved internal passages, cannot be made in one piece by other plastics processing techniques (Fig. 1.4).

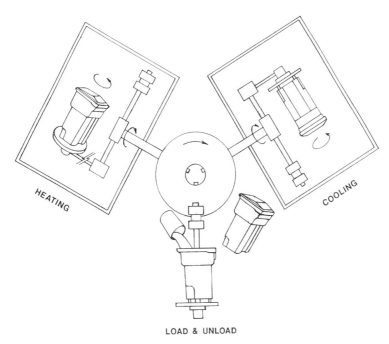

HEATING

COOLING

LOAD & UNLOAD

Figure 1.2 A three-arm turret molding machine that integrates the four steps of the rotational molding process

Figure 1.3 An array of typical parts produced by the rotational molding process (Courtesy Solar Plastics, Minneapolis, MN)

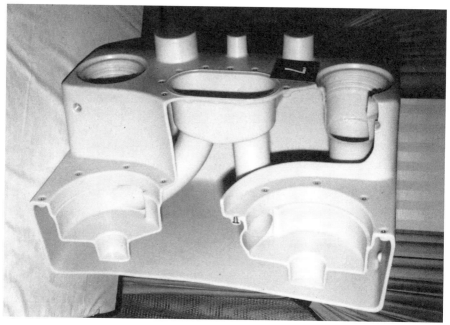

Figure 1.4 A one-piece sludge-lift tank with inside threads, metal inserts, and curved internal passages (Courtesy Rhein-Bonar Kunststoff-Technik GmbH, Hockenheim, Germany)

This is a low-pressure process that allows the use of light-duty molds and machines. The correspondingly lower mold and machine costs allow the production of small quantities of very large or complex parts that would not be economically feasible with other, more capital-intense, high-pressure processes.

The molds required for this process can normally be built and put into production sooner than the tooling required for other plastic-forming processes.

This process uses little or no pressure in comparison to other closed-molding processes that require higher processing pressures to force the material to flow, or form, into the cavity. The heating and cooling of plastic under low-pressure conditions results in molded parts that exhibit low levels of molded-in stress. This reduction in molded-in residual stress results in improved impact strength and chemical resistance, with a reduction in postmold warpage.

There is no pressure inside the mold to force the liquid or powdered plastic into the shape of the cavity. In this case, the hot cavity is rotated to bring it into contact with the plastic material. There is, in effect, little or no flow of the plastic material with this sintering process. This allows the molding of parts with walls that can be extremely thin in relation to their overall size. These thin walls reduce the amount of plastic material required and minimize molding cycle time. The end result is a lower cost part.

The sintering nature of this process allows the formation of parts with wall thicknesses that are more uniform than comparable parts made by the competitive hollow-part processes based on the stretching of a preform.

Rotational molding is the ideal plastics process for running family molds. Widely different shapes of large and small parts, in single and multiple cavities, can be efficiently run simultaneously on a machine.

The ability to quickly change from one mold to another allows the production of the small quantities required for just-in-time deliveries.

The mold's cavity is the only thing that comes into contact with the plastic material. There is no time lost in purging from one kind of material or color to the next. This is a distinct advantage over the melt-flow processes, which can require several hours of purging.

This process has the ability to produce two-color or two-material parts without multiple molds or sophisticated two-material molding machines. Two- and three-layer hollow parts can be solid, foamed, or combinations of both.

Rotational molding is an excellent process for producing parts with molded-in inserts. Metal, plastic, rubber, and wooden inserts in lengths of one meter have been successfully molded into one-piece composite parts.

Some types of parts can be molded without the draft angles that are required for closed molding processes.

The hollow nature of these parts, with their lack of internal cores, allows the molded part to shrink away from the cavity and flex inward to accommodate undercuts without complicating the tooling.

Parts produced by this process are free of weld-lines, gate vestiges, and scars left by ejection mechanisms.

Rotational molding is the best plastics manufacturing process for producing deep-draw, hollow parts with closely spaced parallel walls. Cooler chests, food-insulating containers, and boats with the inner and outer hulls molded as one piece are examples of double-walled parts of this type (Fig. 1.5).

Rotational molding produces a minimal amount of scrap material. There are no sprues or runners as there are in injection molding. The finished parts are not cut out of a larger sheet of material, as is required with the thermoforming process. The pinch-off scrap that is an inherent part of extrusion blow molding is not present. All the material that goes into the mold comes out as a finished part.

Industrial and postconsumer polyethylene can be recycled with a minimum loss in physical properties.

Figure 1.5 The cross-section of a boat showing integrally molded inner and outer hulls (Courtesy Association of Rotational Molders)

1.3.2   Limitations

Like all other plastics processing techniques, rotational molding has its fair share of limitations. On the negative side, the two greatest limitations are that as an open-molding process, there are no cores inside the hollow parts. Surface details and dimensions can only be provided and controlled on the side of the part that is formed in contact with the cavity. Second, the process requires the heating and cooling of not only the plastic material, but of the mold as well. Most other thermoplastic processing techniques heat and cool only the material. Heating and cooling the mold as well as the plastic results in higher energy costs and longer molding cycles.

The long heating cycles and high oven temperatures increase the possibility of thermal degradation of the plastic material being molded. A major limitation to rotational molding is that there are some plastic materials without the required heat resistance to withstand these long heating cycles.

Another plastic material related limitation is that the material must be in a liquid form, or be capable of being effectively pulverized into a fine powder that flows like a liquid. Pulverizing plastic pellets into a powder adds to the material's cost. The plastic material used for the rotational molding process normally costs more than the same material processed by a technique that can use the material in a pellet form.

The plastic part is free to pull away from the cavity during the cooling and shrinking portion of the molding cycle. This situation makes it difficult to maintain precise dimensions and large, flat surfaces that are free of warpage.

Getting the plastic to stick to and sinter onto the surface of the cavity and then release from the cavity during the cooling cycle requires the careful application of mold release. This is an added manufacturing procedure that is not required by some other plastics processing techniques.

1.4   A Brief History

Today, it is difficult to imagine such a thing, but once upon a time, there were no plastic materials and no rotational molding process. The process of rotational molding evolved before manmade plastic materials became available.

The first known, documented use of heat and biaxial rotation for the forming of hollow parts was a British patent, No. 1301, issued to R. Peters in 1855. The Peters patent was directed toward an improvement in the manufacture of metal ordnance shells and other hollow vessels. All the basic elements of the rotational molding process were described, but Peters relied on centrifugal force to push the material into contact with the cavity.

A decade later, in 1865, United States Patent No. 48,022 was granted to T. J. Lovegrove. This patent was also concerned with the production of hollow, round artillery shells of a uniform wall thickness. The claimed advantage of this technology was that the improved uniformity in wall thickness and density resulted in more accurate flight of the projectile. It is amusing to note that the modern rotational molding industry still relies on this process to produce plastic parts of a uniform wall thickness. It is also interesting to note in passing that the Lovegrove patent was issued three years before John Wesley Hyatt's invention of cellulose nitrate, the first manmade plastic material.

The forming of hollow wax objects is described in United States Patent No. 803,799, which was granted to F. A. Voelke in 1905. The molding of hollow chocolate eggs was covered in a 1910 United States Patent, No. 947,405, issued to G. S. Baker and G. W. Perks.

R. J. Powell was granted United States Patent No. 1,341,670 in 1920 for a machine for molding plaster of Paris objects. This landmark patent taught the advantages of slow rotation and the elimination of centrifugal force. This patent also described the still widely used four-to-one ratio of rotation.

The first mention of a *rock-and-roll* rotational molding machine appears in United States Patent No. 1,875,031, issued to H. B. Landau in 1932. This patent described a machine for molding wax bottles.

A machine for producing rubber balls was patented in England by W. Kay in 1932. That invention was cross-patented in the United States as Patent No. 1,998,897, which was issued in 1935. This patent illustrated the biaxial rotation-generating gear drive that is similar to that used on modern machines.

These and other early patents paved the way for the rotational molding of hollow plastic articles. By the time the Kay patent was issued in 1935, the basic elements of rotational molding had been known and practised for eighty years, but the early plastic materials were not suited to the process. Fortunately, the rubber industry continued the development of the machines and the molding procedures, while keeping the process alive. Some plastic parts were produced, but the industry had to wait for another decade before a suitable plastic material became commercially available.

B. F. Goodrich's Waldo Semon is credited with discovering that rigid polyvinyl chloride (PVC) could be softened by adding a plasticizer [1]. Prior to this 1926 discovery, polyvinyl chloride was considered to be of no commercial value. Softening the material paved the way for major applications, such as packaging film, electrical insulation, and rotational molding.

As an aside, Semon also invented bubble gum, which was immediately recognized as having great commercial significance.

The first documented rotational molding of PVC appears in a description of the casting of synthetic liquid syrups, such as styrene, vinyl acetate, methyl methacrylate, and vinyl chlorides. This early teaching is found in United States Patent No. 2,265,226, which was issued in 1941 to J. H. Clewell and R. T. Fields. The patent was assigned to E. I. du Pont de Nemours & Co.

The Union Carbide Company introduced plasticized liquid PVC in 1946. This was the material that rotational molders had been waiting for. These materials, which became generally known as plastisols, were well suited to the process of rotational molding. The first full disclosure of the rotational molding of polyvinyl chloride in a liquid plasticizer appears in a 1948 Italian Patent, No. 440,295, which was issued to C. Delacoste and Y. Cornic.

A noteworthy United States Patent, No. 2,629,134, was issued to R. P. Molitor on February 24, 1953. This patent embraced the teachings of earlier patents and claimed the two-phase heating of plastisol to achieve gelling followed by fusion at a higher temperature.

The Molitor patent was assigned to the Sun Rubber Company. The protection provided by this patent was vigorously prosecuted by this company, and the ensuing patent litigations had a negative effect on the growth of the industry from the mid-1950s through the early 1960s. Plastic material and molding machine manufacturers assumed a wait-and-see attitude until these litigations were resolved. The Molitor patent was eventually ruled invalid for lack of novelty.

The Sun Rubber Company litigations are only interesting from the perspective that they slowed the growth of the industry for almost ten years. Shortly after the company lost their petition for a rehearing of the case, the industry resumed its steady rate of growth.

The material manufacturers responded to this opportunity with an expanded line of products. Physical properties were compounded to range all the way from the stiffness of wood to a softness approaching rubber. The basic plastisol material was transparent, but it could be economically colored and easily decorated. The molding of plastisols grew into an important industry. This versatile material dominated the rotational molding industry all through the 1950s and 1960s. Plastisol was a great material, but it could not satisfy all the marketplace opportunities. There were definite limits to what a molding industry with only one material could do.

As rotational molding grew into a recognizable part of the plastics industry, it caught the attention of the machinery builders. Before this time, many molders built their own molding machines. It was jokingly claimed that you couldn't really be a rotational molder unless you built your own machines. During the 1950s, commercially available rotational molding machines became common. There were seven rotational molding machine builders in the United States in 1954. The E. B. Blue Company marketed batch machines specifically designed for the multiple-cavity molding of plastisol. Akron Presform developed a multi-arm machine. In 1964, the McNeil Company marketed a multi-arm turret machine that was to become the standard in the industry.

An equally important, but unrecognized at the time, machinery development was an efficient powder-pulverizing mill. The first of these machines was produced in Germany by the Pallmann Pulverizers Company, Inc., in 1955 [2]. This new style of pulverizer had the important advantage of being able to produce fine powder from low temperature softening plastic materials without the use of cooling agents.

Mold makers also took note of the emerging rotational molding industry. Electroformed copper had been used since the 1850s as a method of duplicating printing plates, art objects, and phonograph records [3]. The electroformed tool-making industry was well developed by the time plastisols were introduced in

1946. Some of the first molds for products such as dolls' heads, balls, figurines, and artificial fruit were made by electroforming. The introduction of sulfamate nickel in 1950 provided electroformers with a tougher and stronger material for the making of tools and molds.

The use of high-quality cast aluminium tooling for rotational molding was pioneered by Kelch Aluminium Molds' Allen Kelch and Plastic-Cast Mold's Bud LaMont [4]. Working independently, both discovered that casting in plaster and carefully controlling the rate of cooling and shrinkage produced aluminium castings with cavity surfaces that were satisfactory for the rotational molding process. The first plaster cast aluminium molds for rotational molding went into production in 1949 or 1950. Today, cast aluminium cavities are the most frequently used molds for this process.

The fabricated sheet metal molds that would, one day, dominate the production of large tank molds had to await the availability of new, stronger plastic materials.

While the industry waited for a new material breakthrough, resourceful custom molders continued to find additional applications for plastisols. These new uses stretched, and sometimes exceeded, the capability of this material. Plastisol's low heat resistance and lack of stiffness were limiting factors. The search for additional materials began in earnest. That search was at least temporarily ended, not by a new plastic material, but rather by an old material coupled with a new development in pulverizing.

In 1933, scientists at Great Britain's Imperial Chemical Industries (ICI) noted traces of a waxy, white substance on the inside surfaces of a pressure vessel [5]. This was a precursor to the polyethylene (PE) material that would one day dominate the entire plastics industry. The high pressures required for the polymerization of this material made the development of PE a tedious process. The advent of the Second World War in September 1939 also hampered commercial development. At the same time, wartime production of PE was pushed forward as a substitute for rubber electrical insulation, which became scarce as the rubber plantations in the Pacific were overrun by enemy forces.

In order to increase production, ICI licensed the Du Pont Company to produce PE in the United States in 1942. Du Pont started production in May of 1943. Following the end of the Second World War, other licenses were granted and production turned to peacetime products. By the late 1940s, the producers of PE were prospering due to the development of two giant markets: the extrusion of film for packaging applications, and blow-molded bottles.

These early PEs were low- and medium-density materials. High-density polyethylene (HDPE) was not discovered until 1957. The rotational molding industry took note of PE's impressive attributes. However, these materials were commercially available in pellet form only. These large pellets were not suitable for the sintering phase of the process. The same Bud LaMont who pioneered cast aluminium tooling is also credited with having rotationally molded the first PE product in 1953, a Mickey Mouse figurine. The powder was produced by abrading particles from sheet stock using a rotary wire brush. The material molded well, but a more efficient method of producing PE powder had to be found.

The solution to this problem already existed in the pulverizing mill developed by Pallmann in 1955. These mills were capable of producing uniform quality fine powders from low temperature softening plastic materials. A few years passed before these two technologies were brought together, but when that was done, the industry had the material it had been looking for.

Pulverized into a fine powder, PE flowed like a liquid. It sintered onto the cavity in uniform thickness. It could be compounded to withstand the long, hot oven heating cycle. It had a broad processing window and was easy to mold. PE had the necessary chemical resistance to be used for all kinds of storage, shipping, and processing tanks. Best of all, it was relatively low in cost.

The first PE in a powder form suitable for the rotational molding process was marketed in 1961 by U. S. Industrial Chemical (now Equistar Chemicals).

The first ever national conference devoted exclusively to rotational molding was held during the National Plastic Exposition in Chicago, Illinois on November 19, 1963. As an indication of the interest created by this relatively new process, more than 1,000 people (including this author) attended that symposium. Liquid PVC plastisol was still the most frequently molded material, but this meeting featured the advantages of the new low-density polyethylene (LDPE) powders. A second symposium that extolled the virtues of HDPE was held in New York City, in 1966. These symposiums were instrumental in establishing PE as the workhorse of the rotational molding industry.

There was a waiting market for rotationally molded PE products. Custom molders quickly put PE into the applications they had found that could not be satisfied with plastisol. The molding industry prospered, and this encouraged other materials manufacturers to enter the market.

In the early days, injection molding grades of PE were used for rotational molding. As competition among material suppliers grew, they tried to differentiate themselves by marketing many different grades of HDPE and LDPE that were specifically compounded for the unique requirements of the rotational molding process. All through the 1960s, the custom molders and their suppliers enjoyed an expanding market that was increasingly dominated by PE. Once again, the limitations of the available plastic materials became apparent. The industry had prospered by taking on increasingly demanding applications, but customers' appetites had been whetted and they were asking for more. PE was not suitable for contact with some chemical reagents. In spite of this limitation, small and medium-sized tanks and containers of all types had become an important market, but PE was not strong enough for the larger tanks that customers wanted to produce. The search for additional materials was renewed.

Polycarbonate (PC) was the first rotationally moldable material that combined transparency with temperature resistance, impact strength, and good stiffness. This material was discovered in 1957, but it took until 1968 before the first transparent lighting globes were rotationally molded by Formed Plastics, Inc.

Rotationally molded gasoline tanks for motorized vehicles debuted with a six-gallon HDPE tank on the Sno-Sports snow sled in the mid-1960s. The transportation industry was recognized as a large potential market for gasoline tanks. However, once again, a better material was required. Phillips Chemical Co. responded to this opportunity with cross-linked polyethylene (XLPE) in 1970. Cross-linking the PE molecules improved the material's chemical stress crack resistance, stiffness, and low temperature impact strength.

By the mid-1970s, the processing part of the rotational molding industry was prospering. This was due in part to this process's ability to produce parts that could not be made as one piece by competitive molding techniques. Back in those days, the molding industry was made up of strong individuals with an entrepreneurial spirit. These were the pioneers of the industry. The process that they practiced was more art than science. There were few technical conferences, and little or no published technical information. Each molder learned by trial and error. These entrepreneurs did not ask for help and they did not share what they had learned, because anyone who was interested was a competitor. This inefficient method of working came to an end in 1976 with the forming of the Association of Rotational Molders (ARM) [6]. This trade association quickly became the premier organization representing the rotational molding industry on a global basis, with members in sixty countries.

Glass fiber reinforced PE and nylon were introduced in the late 1970s. The anticipated markets did not materialize and these materials were withdrawn. Fortunately, the technology was developed and is still available.

A moldable grade of Nylon 6 was marketed in 1978. Nylon's stiffness, high heat deflection temperature, and resistance to hydrocarbons allowed rotational molders to penetrate additional markets, such as fuel tanks and high temperature heating ducts for the transportation industry.

In 1979, the rotational molding industry was the beneficiary of a stroke of good fortune with Prof. Roy Crawford's establishment of the Rotational Molding Research Centre at Queen's University of Belfast, Northern Ireland. The ongoing research work at Queen's University has greatly contributed to the understanding of the technology that controls the rotational molding process. Of equal or perhaps greater importance are Prof. Crawford's students. After finishing their graduate studies at the Research Centre, these students spread out through the industry. It can be anticipated that these rotational molding professionals will continue to make breakthroughs in rotational molding technology for years to come. In fact, the results of their efforts are already being realized.

Following Queen's University's lead, other teaching and research institutions around the world started offering rotational molding programs.

The 1980s were a great decade for the rotational molding industry. One important event was Du Pont's introduction of linear low-density polyethylene (LLDPE). This material, which had been available since 1964 in Canada, was introduced into the U.S. market in 1980. These easy-to-process materials provided a combination of properties that rivaled XLPE for heat, chemical, and low-temperature impact resistance. Linear PEs were quickly accepted and grew to be the most common rotationally molded material.

In another landmark development, the flexibility and repeatability of molding conditions were greatly improved with the commercial availability of microprocessor-controlled molding machines.

Rototron Corp. started selling rotational molding grades of polypropylene (PP) in 1975, but this material was handicapped by its low impact strength. A. Schulman and Equistar experimented with PP in the early 1980s. All three companies were offering commercially successful PPs by 1995.

The first of the metallocene catalyst system patents, No. 4,530,914, was issued in 1985. The inventors were J. A. Ewen and H. C. Welborn, Jr. Patent rights were assigned to Exxon Research and Engineering Co. This patent would lead to the single-site polymerization processes that would allow the production of PE and other polymers, which would benefit the plastic material and processing industries. The first of these materials for rotational molding became available in the mid-1990s. Early indications are that these materials will prove to be very useful in the future.

Manufacturing processes are usually developed to adapt existing materials to take advantage of a marketplace opportunity or problem. Rotational molding is an exception. The process existed before plastic materials became available. In spite of this exception, the history of the rotational molding industry can be traced to the availability of suitable materials, starting with iron ordnance shells. Wax, chocolate, plaster of Paris, and rubber all had their moment of glory, but today and tomorrow belong to plastics. As additional new plastic materials become available, this industry will find uses for them.

1.5   The Business Today

The modern rotational molding business is well established and growing. Most major processors and their suppliers have come together under the auspices of ARM. ARM has now become the international focal point for many mutually beneficial activities. The primary goals of the organization are to promote the industry, educate its members, and refine the technology. By combining resources, the industry is able to carry on an international promotional campaign. ARM's technical committees continue to refine and publish technical information. This work is backed upon by university research projects sponsored by the association. All of this technical information is disseminated through *Rotation* magazine [7], expositions and conferences in the U.S., Europe, and elsewhere.

The University of Wisconsin at Milwaukee [8] started sponsoring *Advances in Rotational Molding* seminars in 1995. This seminar is structured to appeal to original equipment manufacturers (OEMs) who want to increase their understanding of this process.

In 1996, the international Society of Plastics Engineers (SPE) [9] started a new division devoted exclusively to serving the needs of the scientific, academic, and end-user segments of this industry. This new SPE activity is being led by Profs. Lorraine Olson and George Gogos at the University of Nebraska at Lincoln, and the author. It can be anticipated that in the future SPE's Rotational Molding Division will become an important resource for the dissemination of technical information relating to this process.

All through the 1980s and the first half of the 1990s, the industry enjoyed annual growth of 10 to 15%. In the late 1990s, this rate of growth declined to 8.5 to 10%. This was due, in part, to changes in the merchandising of large toys and playground equipment. Some hypothesize that the rotational molding industry has topped out, but that is not the case. The industry is definitely maturing, but it has not yet reached its full potential. What is actually happening is that the industry has paused to catch its breath, regain its strength, and consolidate its position for the next big growth phase. Toys and playground equipment are only two of this industry's many markets. This industry has a proven record of finding new outlets for its products. It can be anticipated that the industry will, once again, find new markets to replace those that have been lost. When the large toy and playground equipment markets return, the industry will be ready. In the meantime, the growth of the rotational molding business is still faster than the rest of the plastics industry.

The exact number of rotational molders in North America is not known. It is estimated that there are approximately 400. This is a relatively small business, compared to the larger injection molding and extrusion industries. Rotational molding has not been analyzed and studied the way the larger processing segments of the industry have been. That situation is now changing, due to the attention generated by the growth of the industry.

The amount of plastic material consumed by the North American rotational molding industry was estimated at 5,445 t (12,000,000 lb) in 1965. A more in-depth 1968 analysis increased that number to 68,000 t (150,000,000 lb). By 1982, consumption had doubled. In 1991, usage reached 181,400 t (400,000,000 lb).

The first professionally done in-depth market survey of the industry was carried out by Plastics Custom Research Services' Dr. Peter Mooney [10]. He estimated the 1994 market at 343,800 t (756,000,000 lb), and in 1996, at 388,700 t (855,000,000 lb). The rapid increases since 1982 are attributed to the thoroughness of these studies, plus the rapid growth of the industry.

A 1996 survey indicated that the major segments of the industry were

Toys	41.0%	Materials Handling	3.6%
Tanks	19.1%	Housewares & Consumer	2.6%
Containers	9.0%	Playground Equipment	2.3%
Automotive	8.7%	Miscellaneous	13.7%

Now, in the late 1990s, the rotational molding business is benefiting from a stable and robust economy. The technological advances in materials, tooling, and processing have refined the industry. What most progressive molders are now doing is more of a science and less of an art.

New rotational molding companies are joining the industry. Smaller and less progressive molders are being acquired by larger, more aggressive companies. Many processors have multiple locations, and some have become global in scope. This contraction, or weeding out, of the industry is resulting in larger and stronger full-service molders, who are well equipped and staffed to meet the increasingly demanding challenges of the twenty-first century. The rotational molding industry is well established and vigorous. It can be anticipated that this industry will continue to provide both business and career opportunities through the foreseeable future.

2 Rotational Molding Materials

2.1 Process-Related Material Requirements

Rotational molding is a materials-dependent process. This manufacturing technique could not exist without suitable plastic materials. The attributes of the process impose certain limitations on the materials that can be molded. In this regard, the process is no different than reaction injection molding, which is at its best with polyurethane. Lay-up, spray-up, and resin transfer molding cannot accommodate thermoplastic materials.

This process differs from other thermoplastic processes in that both the plastic material and the mold must be heated and cooled during each molding cycle (Fig. 2.1). This results in relatively long cycle times. The high heat and long cycle times provide an opportunity for thermal degradation of the plastic material.

During the heating portion of the molding cycle, the hot plastic is in contact with the oxygen in the air that is trapped inside the cavity. This can lead to a loss in physical properties, due to oxidation of the hot plastic material. Those plastics that are specifically compounded for rotational molding contain more than the average amount of antioxidant to protect the material from oxidation.

There are no forces on the plastic material to push or pull it into contact with the cavity. In the rotational molding process, the material remains as a puddle or pool of plastic in the bottom of the cavity (Fig. 2.1). As the machine rotates the mold through two axes, all surfaces of the cavity repeatedly pass through the puddle of plastic material. Molding is actually achieved by the powdered or liquid plastic material adhering to, or sintering onto, the cavity. In order for this process to work in the absence of pressure, the individual powdered particles of material must be capable of flowing together to produce a homogeneous, solid wall. This *no-pressure* molding process requires materials with a melt index of at least three for simple shapes, such as balls and tanks. Complex-shaped and multifunctional industrial parts with fine surface details require a melt index of five or greater.

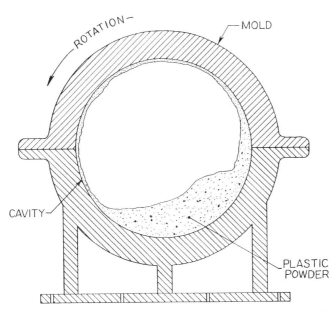

Figure 2.1 During the rotational molding process the mold rotates while the plastic material remains in the bottom of the cavity. The plastic adheres to the hot cavity as it passes through the material

The rotational molding process is limited to those polymers that are available as a liquid, or are capable of being efficiently pulverized into fine powders that will flow like a liquid. There is a limited number of liquid plastic materials. Some materials cannot be economically pulverized into uniform powder particles. Some work is now being done with micropellets. The initial results are encouraging, but the suitability of micropellets for general use is just now being determined.

2.2   Common Plastics

The three process-related material requirements of 1. availability in a powder or liquid form; 2. thermal stability; and 3. good flow, combine to eliminate many plastic materials from consideration. Five of the plastic materials that overcame these limitations are listed in Table 2.1, in descending order of usage. These are the commonly rotationally molded materials.

Table 2.1 The Most Commonly Rotationally Molded Plastics and Their Chemical Structure

Plastics	Structure
Polyethylene	$\left[\begin{array}{cc} H & H \\ \vert & \vert \\ C - C \\ \vert & \vert \\ H & H \end{array}\right]_n$
	LLDPE LDPE HDPE XLPE
Polypropylene	$\left[\begin{array}{cc} H & H \\ \vert & \vert \\ C - C \\ \vert & \vert \\ H & CH_3 \end{array}\right]_n$
Polyvinyl Chloride	$\left[\begin{array}{cc} H & H \\ \vert & \vert \\ C - C \\ \vert & \vert \\ H & Cl \end{array}\right]_n$
Nylon	$\left[\begin{array}{cccccc} H & H & H & H & H & O \\ \vert & \vert & \vert & \vert & \vert & \vert\vert \\ N-C-C-C-C-C-C \\ \vert & \vert & \vert & \vert & \vert \\ H & H & H & H & H \end{array}\right]_n$ Nylon 6 Polycaprolactom
Polycarbonate	$\left[O-C\begin{array}{c}H\ H\\C=C\\C=C\\H\ H\end{array}C\begin{array}{c}CH_3\\ \\CH_3\end{array}C\begin{array}{c}H\ H\\C=C\\C=C\\H\ H\end{array}C-O-C\!\!=\!\!O\right]_n$

2.3   **Other Plastics**

In addition to these five industry leaders, there are other plastics that are also molded in some instances. These other materials are shown in Table 2.2. In most cases, the limited usage of these other materials has not justified the development of grades that are compounded for the special requirements of the rotational molding process. The use of these other materials is also hampered by the limited number of processors who have learned how to mold them.

Over the years, some of these other materials have been gradually changed to suit the needs of other processing techniques with larger market opportunities. For example, acrylonitrile butadiene styrene (ABS) was being rotationally molded in the mid-1970s. Today, there are no commercially available ABS materials suitable for the process. ABS is only molded in rare cases, using proprietary compounding and molding procedures.

Among the other materials, there are several that are worthy of comment, because they can be rotationally molded and because their usage is increasing.

The four members of the large fluorocarbon family that are now being molded are ethylene-chlorotrifluoroethylene (ECTFE), ethylene-tetrafluoro ethylene (ETFE), polyvinylidene fluoride (PVDF), and perfluoroalkoxy (PFA). The ETFE and PFA polymers are specified most often. They are finding increasing use in the semiconductor and chemical-processing industries as tanks and processing containers, and as linings inside metal tanks, valves, and pipes. Both materials are crystalline, branched copolymers that flow well enough to be rotationally molded. These fluorocarbons are specified for their combination of extremely good chemical resistance and nonstick, easy-to-clean surfaces, and their high operation

Table 2.2 Other Rotationally Molded Plastics

Acrylonitrile butadiene styrene
Acetal
Acrylic
Cellulosics
Epoxy
Fluorocarbons
Ionomer
Phenolic
Polybutylene
Polyester
Polystyrene
Polyurethane
Silicone

temperatures. At low loads, the ETFEs can be used continuously at 149°C (300°F), and the PFAs go up to 260°C (500°F). Between the two, PFA has better chemical resistance. ETFE is known for bonding securely to metal surfaces.

Some of the thermoplastic polyesters are now being rotationally molded. These are the elastomeric block copolymer materials based on polyethylene terephthalate. These materials have a good combination of impact strength and chemical and heat resistance. They are usable at low loading from −40 to 149°C (−40 to 300°F). The polyesters are specified where flexible LDPE and plastisols do not have enough heat resistance. The major markets to date are as flexible bellows and boots in under-the-hood automobile applications, where they are in contact with heat, gasoline, oil, and grease. Liners for compressed natural gas fuel tanks are an application now being considered.

Commercially available liquid polyurethanes and silicones are both rotationally moldable. These two thermosetting materials cross-link during the molding process. The polyurethanes find their widest usage in applications requiring low-temperature impact strength, coupled with abrasion resistance. The silicones are known for their flexibility and extremely high service temperature. Mineral-filled grades can be used continuously at temperatures of 149 to 260°C (300 to 500°F). Their primary uses to date have been as chemically resistant nonstick surfaces, and as prostheses such as a lifelike human hand.

These other materials are important to the companies that have a need for them but, as a group, their usage accounts for less than 2% of the total market by weight. The physical properties and sources for these other materials can be found in the *Modern Plastics Encyclopedia* and the *International Plastics Selector.*

2.4   Special Plastics

The common and other rotationally moldable plastic materials are specifically compounded to provide features that are not present in a single material. Specialty compounding is a growing field. In this regard, rotational molding is the same as other plastics processing techniques.

2.4.1   Reinforced Plastics

Some successful work has been done with fiber-reinforced PE and nylon. Adding long-fiber reinforcements to the thirty-five mesh powder does not result in a homogeneous molded part. The powdered plastic adheres to the cavity, while the fibers tend to migrate to the inside of the part. When a homogeneous part is produced, the physical properties are improved, as expected. The one exception is that there is a reduction in impact strength. The presence of the fiber can also hinder flow and cause surface imperfections. Currently, there are no commercially available fiber-reinforced rotationally moldable plastic materials.

2.4.2   Filled Plastics

Low-cost fillers or extenders are being successfully used to reduce the amount and cost of the plastic required to produce a molded part. With the proper filler, there is a minimal loss in physical properties. These combinations of molding materials and fillers must be carefully selected and compounded to optimize the results without a significant loss in impact strength. This technology is now being given increased attention, and improvements can be anticipated.

2.4.3   Multiwall Parts

It is a common rotational molding practice to combine two different colors of the same material, or two dissimilar materials, into one part. A part with one color on the outside and a second color on the inside can be used for decorative purposes.

Two dissimilar plastics can be molded as one part to provide functional features that cannot be achieved with one material (Fig. 2.2A). A thin layer of uncolored nylon, with excellent resistance to gasoline, could be inside a colored PE tank. In that case, the lower cost PE would provide structural strength and outdoor weatherability. Another application could be a storage tank molded in reprocessed PE, with an inner layer of virgin PE, that has National Sanitation Foundation certification for potable water.

The double-wall process is at its best when the two walls adhere to each other. Two different colors, or virgin and reprocessed combinations of the same material, would be ideal. Dissimilar materials, such as nylon and PE that do not bond to each other are being used, but there are some limitations. The two materials must be chemically compatible. They should have similar processing temperatures and similar coefficients of thermal expansion.

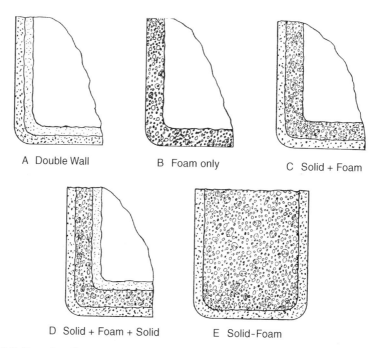

A Double Wall   B Foam only   C Solid + Foam

D Solid + Foam + Solid   E Solid-Foam

Figure 2.2 Rotationally molded parts can have walls with A) two unfoamed, solid materials, B) one foamed material, C) an outer solid wall with a foamed inner wall, D) solid inner and outer walls with foam in between, E) a solid outer wall completely filled with foam

2.4.4   Foamed Parts

Structural foam parts have been produced by rotational molding for many years. The process can produce parts with walls that are basically all foam, with little or no solid skin (Fig. 2.2B). A more common approach is a composite wall with a solid skin on the outside, and a layer of foam on the inside (Fig. 2.2C). Parts of this type use the foam for sound or heat insulation, added stiffness, or to achieve a reduction in the amount of material being used. Properly done, the solid skins are free of the swirls and silver streaking that are common with injection molded structural foam.

A more sophisticated approach is a three-layer composite wall, with the foam sandwiched between two solid layers (Fig. 2.2D). This approach provides all the advantages of foam, but with a solid, dense surface on both sides of the wall. Manufacturers of camper tops, furniture, and boats have used this technique to great advantage.

Another approach is to mold foam into the open space between two closely spaced parallel walls. Ice chests, floating docks, cart doors, and panels are now being produced this way (Fig. 2.2E).

The most commonly used technique for producing multiwall and foamed parts is by the multicharge process. With this procedure, the first charge of plastic, which will become the outside surface of the part, is charged into the mold in the normal manner. The second charge of material that will form the inner surface of the part is charged into an insulated chamber or *drop box* that is attached to the mold. This second charge of material may be a different color, a different material, or a material that contains a foaming agent. The insulated drop box opens and charges the second material into the mold only after the first charge has been sintered onto the cavity.

Molders have been producing foamed parts since the early 1970s, using the single-charge method. With this technique, the plastic is a blend of lower and higher temperature melting materials, or a blend of smaller and larger powder particle sizes. The process works on the principle that the outer surface of the part will be formed by the lower temperature or smaller particles that melt first and sinter onto the cavity before the higher temperature or larger particles which contain the foaming agent, have time to melt.

Single-charge foaming materials specifically compounded for rotational molding became commercially available from Equistar and Wedtech in the mid-1990s. Single-charge foaming is a new technology, with all the negative implications that implies. Research continues, however, and singe-charge foaming will become the choice for foamed parts in the future.

Capitalizing on plastic's corrosive and chemical resistance, special plastics are being specifically compounded for special applications. Today, there are commercially available special materials for lining metal tanks, pipes, and valves for use in the manufacturing of corrosive chemicals or abrasive materials, such as cement. The medical, food-handling, and semiconductor industries make wide usage of this technology.

Other special plastics are compounded to provide unusual decorative effects, including the popular speckled granite look. With the continued effort to improve PE, it was discovered that by adding a comonomer of butene, hexene, and octene to the reactor, it was possible to produce polymers with special characteristics. In general, there was an increase in cost and properties with the progression from butene to hexene to octene. Examples of the effects of these three comonomer PEs are shown in Table 2.3.

Other special materials are created by combining standard plastics with modifiers.

Table 2.3 Examples of Comonomer-Modified

	Butene	Hexene	Octene
Impact strength HDPE 0.125 in. Thickness @ −40°C Approx. ftlb to failure	40–45	50–55	55–60
Environmental stress-crack resistance LLDPE 10% igepal. approx. hrs. to failure	50	500	750
Cost	lowest	average	highest
Average ± cost/lb.	−$0.02	average	+ $0.04

2.5   Additives and Modifiers

Additives are normally mechanical mixtures of a plastic material with one or more special modifiers that provide a property that the plastic material does not possess. The rotationally moldable common, other, and special materials can all be specially compounded and modified with additives, just like the materials used for other processes. The most commonly used additives in the rotational molding industry are as follows:

1. Pigments and dyes are added to plastic material to produce a colored part. Colorants are one of the most commonly used additives.
2. Foaming agents are material that thermally decomposes to form bubbles or foam in the plastic material.
3. Toughening agents are low molecular weight or elastomeric materials that distribute applied stresses over a broader area to resist a rapidly applied load.
4. Flow enhancers are used to improve the flow of the plastic material between individual powder particles, and into fine or difficult to mold cavity details.
5. Internal lubricants are added to a plastic material to ensure that the part releases from the cavity. Their use may minimize or eliminate the use of an external lubricant.
6. Antistatic additives are used to eliminate the buildup of a static charge as the metal cavity rotates through the powdered material. They are also used to prevent the separation of powdered pigments from the plastic material. In some cases, they improve the flow of the material.
7. Cross-linking additives are added to polyethylene to cross-link the material and increase its environmental stress crack resistance (ESCR) and strength.

8. Antioxidants are used to increase a material's resistance to oxidation with the oxygen in the mold during the heating portion of the cycle. These heat stabilizers protect the material from thermal degradation.
9. Ultraviolet light stabilizers can improve a plastic material's resistance to the deleterious effects of exposure to sunlight.
10. Fire retardants reduce the rate at which a plastic material burns.

All the other special and modified plastic materials have a place in this industry. The success of any rotational molding project will, however, be enhanced by working with one of the five common materials.

2.6   Polyethylene

PE is the workhorse of the rotational molding industry. PE accounts for approximately 85% of the material being rotationally molded. PE is actually not a specific material, but a whole family of materials. The commonly molded members of this versatile family of plastics are LLDPE, LDPE, MDPE, HDPE, XLPE and ethyl vinyl acetate (EVA).

2.6.1   Low-Density Polyethylene

In 1961, Equistar introduced LDPE as a thirty-five mesh powder. This is the oldest member of the PE family to be rotationally molded. As a result, the material and molding technology is well developed.

LDPE, like all other thermoplastic materials, is made up of long molecular chains. These long chains are composed of repeating groups of molecules, which are connected together to create polymers. LDPE contains only carbon and hydrogen atoms (Table 2.4). The LDPE molecule is characterized by having many side branches. LDPE derives its name from the fact that these side branches prevent the individual molecules in a polymer from packing tightly together. This loose packing of the molecules produces a low-density material. Any PE with a density in the range of 0.910 to 0.925 gm/cm^3 is an LDPE.

Table 2.4 Summarizing PE Properties

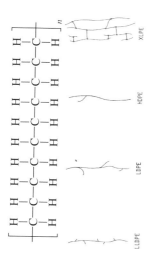

Properties	Standard	Unit	LLDPE	LDPE	HDPE	XLPE
Tensile yield	D-638	MPa	13.1...20.7	7.9...11.0	20.7...27.6	17.9...24.1
		psi	1,900...3,000	1,140...1,600	3,000...4,000	2,600...3,500
Flexural modulus	D-790	MPa	427.4...854.9	138.6...365.4	930.7...1447.7	482.6...792.8
		psi	62,000...124,000	20,100...53,000	135,000...210,000	70,000...115,000
Heat-deflection temperature	D-648	°C @ 0.455 MPa	87.8...97.8	23.0...−12.0	67.2...71.7	65.0...66.1
		°F @ 264 psi	108...122	N.A.	107...111	115...152
Izod impact (notched)	D-256	J/m	774.0...N.B.	389.7...N.B.	245.5...389.7	53.4...1067.6
		ft.lb./in.	14.5...N.B.	7.3...N.B.	4.6...7.3	1.0...20.0
Low-temperature impact	ARM (0.635 cm)	J (−40°C)	33.9...81.3	81.3	135.6...183.0	94.9...135.6+
		ft.lb. (−40°F)	25...60	40	100...135	70...100+
Density	D-1505	g/cm³	0.926...0.940	0.910...0.925	0.941...0.959	0.933...0.945
Mold shrinkage	D-955	%	2.0...2.2	1.5...2.0	1.5...3.0	2.0...3.0
Melt index	D-1238	g/10 min	2.5...14.3	4.5...22.0	5.0...8.0	N.A.

N.A. = Not applicable or not available

N.B. = No break

2.6.2   High-Density Polyethylene

A PE with a density in the range of 0.941 to 0.959 gm/cm^3 is considered to be an HDPE. HDPE is also composed of carbon and hydrogen atoms, but it is approximately fifty times longer than the LDPE molecule. HDPE also has only about one-tenth as many side branches as LDPE. The reduced number of side branches allows HDPE molecules to pack tightly together, and the density to increase accordingly.

There are no chemical or physical forces holding the individual molecules together in a polymer. There is, however, electrostatic attraction between molecules, known as *Van der Waal's Forces*. These forces act like electromagnetic forces, in that the attraction becomes stronger as the distance between molecules is reduced. Straight, symmetrical molecules with few side branches, such as HDPE, can arrange themselves close to one another. The intermolecular Van der Waal's Forces will be high, and the material will exhibit high rigidity. Highly branched, nonsymmetrical molecules, such as LDPE, cannot pack as tightly together. In this case, the intermolecular forces will be lower. The molecules will be able to move more freely and the resulting polymer will be less rigid.

To digress, these intermolecular forces are affected by heat. The higher the heat, the weaker the Van der Waal's Forces become. This explains why plastics lose their strength at elevated temperatures, while becoming stronger at reduced temperatures.

Heating the plastic to a high enough temperature allows the molecules to easily move relative to each other. This allows plastics to be molded into new shapes. The part is held in the new shape until it cools, and then the intermolecular forces return and the part stays in the new shape. This is the principle on which all thermoplastic molding processes are based.

Another important difference between branched and nonbranched molecules is molecular entanglement. This is the characteristic that gives the highly branched LDPE its toughness or *notched Izod* impact strength.

Another effect of the close packing of molecules is the formation of areas with a definite pattern to their molecular interrelationship. Plastic materials containing crystals of this type are referred to as *crystalline*. A polymer with little or no crystallinity is referred to as *amorphous*. Most polymers are a mixture of crystalline and amorphous areas.

The common rotational molding materials that are known for their crystallinity are HDPE, PP, and nylon. A properly molded HDPE part will be 70 to 90% crystalline. LDPE will contain 45 to 65% crystallinity. Some basically amorphous materials are PVC and polycarbonate (PC).

The degree of crystallinity of a plastic material has a significant effect on the material's physical properties. The degree of crystallinity can be changed by the way the material is processed. All PEs are amorphous in the molten state. If the polymer is allowed to cool slowly, the crystals can reform. If the polymer is cooled quickly, the crystals do not have time to reform, and the polymer will be more amorphous. Drastic changes in the cooling conditions can have a significant effect on the degree of crystallinity and the physical properties of the molded part.

Generally speaking, as the degree of crystallinity increases, there will be a corresponding increase in the material's density, tensile and flexural strength, mold shrinkage factor, and heat and chemical resistance. Impact strength and transparency will decrease. Mold shrinkage will become less uniform.

2.6.3   Medium-Density Polyethylene

A PE with a density between 0.925 and 0.941 gm/cm^3 is considered an MDPE. This polymer's shape, degree of crystallinity, and physical properties are midway between LDPE and HDPE. In the past, MDPE was an important rotational molding material. However, it now accounts for an insignificant part of the market.

2.6.4   Linear Low-Density Polyethylene

The commercial arrival of LLDPE in 1980 was a major improvement over the commonly used LDPE. The LLDPE molecule is similar to LDPE, except that the side branches are shorter and are arranged more uniformly along the length of the molecule. This allows these molecules to retain their low density, while packing more tightly together than LDPE molecules. The improved physical properties allowed LLDPE to be used in applications that had previously only been satisfied by HDPE and XLPE.

The advantages of LLDPE were quickly recognized, and this material rapidly became the largest volume in PE being rotationally molded. In the majority of cases, LLDPE has now replaced LDPE in rotational molding applications.

2.6.5   Cross-Linked Polyethylene

In addition to the branched and nonbranched materials, the industry has developed XLPE. In this case, the molecules arrange themselves in a specific pattern relative to each other, with an actual chemical bond being formed between the individual molecules. This, in effect, locks the molecules into a rigid three-dimensional shape. Heating these materials does not weaken the bonds between the molecules. As a result, the cross-linked materials retain their physical properties at elevated temperatures.

The primary advantages achieved through cross-linking are improved creep, low temperature impact strength, and heat and chemical resistance, with a reduction in permeation. These attributes allow XLPE to be used in demanding applications such as fuel tanks, and as self-priming, chemically resistant pump bodies (Fig. 2.3).

Cross-linking of the PE molecules is most often achieved by the addition of 1.5 to 2.0% of dicumyl peroxide. Cross-linking takes place at about 182°C (360°F). During the molding process, care must be taken to make certain that all the plastic has adhered to the cavity before the molten polymer reaches the cross-linking temperature. The material will not flow after it has cross-linked.

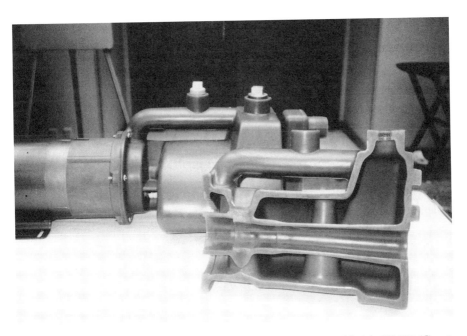

Figure 2.3 A chemically resistant, self-priming pump body molded in XLPE (Courtesy Association of Rotational Molders)

Regrettably, once these materials are cross-linked, they cannot be remelted and reprocessed by rotational molding.

Recent research conducted by the University of Massachusetts at Lowell on a grant from ARM indicates that XLPE can be reprocessed by high shear rate processes, such as injection molding.

2.6.6   Ethyl Vinyl Acetate

EVAs are copolymers of PE and vinyl acetate. These are soft, flexible materials with a heat deflection temperature of only 37°C (99°F). They are more transparent and have a higher cold temperature impact strength than the other PEs. EVAs are frequently blended with other PEs to produce a softer or higher impact type of material.

This material competes directly with LDPE and the stiffer grades of PVC. Typical applications include soft, squeezable toys, traffic cones, ice sculpture molds, energy-absorbing highway barricade covers (Fig. 2.4), airport runway light encasements, and bumpers for boats and bumper cars.

Figure 2.4 Impact-resistant, EVA highway barricade end cover (Courtesy Association of Rotational Molders)

2.6.7   Summarizing Polyethylene

There are many different members of the PE family of materials. The most important physical properties of the PEs are shown in Table 2.4, along with an illustration representing the repeating portion of the different molecules at Fig. 2.1. In addition to the primary type of PE, there are also many specific grades within each type. The listed properties in this table and those that follow are the highest and lowest values for each type of PE that is currently being offered for sale. All values are from tests performed on rotationally molded samples, unless otherwise indicated.

Each member of the PE family is different, but they do share many common characteristics. On the positive side, all PEs are low in cost and light in weight. They have good impact strength and excellent ESCR. Many PEs meet FDA and NSF requirements. A full range of colors is available.

The limitations of PE, in comparison with the other commonly molded plastics, are its low stiffness and temperature resistance. PEs are difficult to modify to achieve weatherability and fire retardancy without a loss of impact strength. All PEs have a high mold shrinkage factor and are prone to warpage. No transparent grades are available. The added cost of grinding PE into a free-flowing, thirty-five mesh powder increases its cost. All PEs are difficult to paint.

It is a common practice in this industry to refer to and classify the many different types of PE by their melt index and density. Density, as determined by ASTM D-792, refers to the weight of one cubic centimeter of the material. The PEs with few side branches can pack tightly together, which encourages crystallinity. As the crystals coalesce, they increase the density of the material. Generally speaking, as the density of a PE increases, tensile strength, stiffness, heat deflection, hardness, and shrinkage will increase. Chemical resistance and permeation will improve. Elongation, impact strength, ESCR, and flow rate will decrease. Weatherability will be unaffected [11].

Melt index, as determined by ASTM D-1238, is a measure of a material's ability to flow through a restricted orifice under prescribed temperature and pressure conditions. An increase in melt index or flow rate results in a decline in tensile strength, elongation, and impact strength. Chemical resistance, weatherability, and ESCR will be reduced. Hardness, stiffness, and permeation will be unaffected [11].

All the different kinds of molding machines and molds can be used to process PE.

2.6.8   Applications

Rotationally molded PE components have now captured applications in many markets, including the following:

- Industrial products: tanks (Fig. 2.5), drums, containers, tote bins, nesting pallets, floor maintenance machine components and tanks, medical carts, video game housings, newspaper and magazine vending machines, tool chests, shipping cases, refuse containers, pump bodies, septic tanks, sewage aerators and lifting units, machine instrument and appliance housings, mannequins, point-of-sale storage and display racks, temporary shelters, utility sheds, salad bars, air ducts, fishnet and navigational floats, portable toilets, grass catchers, and composting containers.
- Transportation products: truck-bed boxes, liners, and tool boxes, ventilating ducts, fuel tanks, crash-test dummies; battery cases, motorcycle saddle bags and fairings; go-kart bodies; highway signs; barricades, bumpers; and truck brush and animal guards (Fig. 2.6).
- Consumer products: infant and adult furniture such as tables, chairs, beds, desks, and benches; insulated food and beverage containers;

Figure 2.5 An array of stationary and portable PE tanks (Courtesy Association of Rotational Molders)

Figure 2.6 PE brush and animal guard, 75% lighter than steel (Courtesy Association of Rotational Molders)

    decorative planters, urns, vases, and pedestals, mail boxes; and doghouses.
- Recreational products: ride-on and ride-in toys; wading pools; sand boxes; playground equipment, row, motor, and sail boats; kayaks and canoes, playhouses, bowling equipment and furniture, toy storage boxes; jogger's strollers; carousel and hobby horses, camper tops; exercise equipment; and balls of all types and sizes.
- Agricultural products: water tanks; animal feeding and watering troughs; tractor cabs, harvesting-machine panels, and grain bins; and fuel, fertilizer, and pesticide tanks.

2.7   Polypropylene

At a glance, the repeating part of the PP molecule (Table 2.5) looks like LDPE. Both monomers are made up of only carbon and hydrogen atoms. The difference is that the PP molecule has side branches arranged in an orderly manner. This

Table 2.5 Summarizing PP

$$\left[\begin{array}{cccc} H & H & H & H \\ | & | & | & | \\ -C & -C & -C & -C- \\ | & | & | & | \\ H & | & H & | \\ & H-C-H & & H-C-H \\ & | & & | \\ & H & & H \end{array} \right]_n$$

Properties	Standard	Unit	PP
Tensile yield	D-638	MPa	19.2 ... 27.6
		psi	2,790 ... 4,000
Flexural modulus	D-790	MPa	1206.5 ... 1365.5
		psi	175,000 ... 198,000
Heat-deflection temperature	D-648	°C @ .455 MPa	68.9 ... 97.8
		°F @ 264 psi	113 ... 133
Izod impact (notched)	D-256	J/m	26.7 ... 117.4
		ft.lb./in.	0.5 ... 2.2
Low-temperature impact	ARM	−40°C J	27.1 ... 61.0
		−40°F ft.lb.	20 ... 45
Density	D-1505	g/cm^3	0.890 ... 0.910
Mold shrinkage	D-955	%	1.0 ... 2.5
Melt index	D-1238		5.0 ... 20.0

allows the molecules to pack tightly together and the Van der Waal's Forces to be high. Properly molded PP will be 50 to 60% crystalline. In spite of the anomalies, PP has a low density of 0.890 to 0.910 gm/cm^3 On the average, PP exhibits higher stiffness and temperature resistance than PE.

The physical properties of PP are summarized in Table 2.5.

PP is a large-volume plastic material used for many processes other than rotational molding. PPs have been rotationally molded from time to time, but those parts have always suffered from a loss of impact strength. Materials with usable impact strength became commercially available from Equistar, Rototron, and Schulman in the mid-1990s. It can be anticipated that PP will become an important rotational molding material as more research is done and more processing experience is gained.

All the conventional molding machines can be used to process PP. This material has a narrow processing window. Overheating will result in a loss of impact strength. Crystallization takes place slowly and PP must be cooled slowly in order to achieve its maximum physical properties.

All types of molds can be used, as long as they have tight parting lines and smoothly drafted surfaces.

PP pellets are too tough to be ground into uniform, good quality powder. Chilling the pellets make them brittle. These cold pellets can be broken into good quality powder. Cryogenic grinding increases the material's cost by 77¢ to 88¢ per kilogram (35¢ to 40¢ per pound).

2.7.1   Applications

PP is specified for many of the same applications as PE. PP's higher stiffness and increased temperature resistance allow it to perform in applications where PE isn't quite good enough. One example is as steam-autoclavable, portable tissue culture chambers, as a replacement for heavier and more costly stainless steel (Fig. 2.7). This application takes advantage of PP's chemical resistance, stiffness, and heat deflection temperature.

Other applications include large chemical shipping drums, radioactive material containers, and high-temperature air ducts. Some work has been done in lining pipes and valves.

PP is relatively new as a rotational molding material. Additional applications will develop as more processors become familiar with this material.

2.8   Polyvinyl Chloride

The first plastic material to be rotationally molded on a commercial basis was PVC. This was the material that introduced this process to the plastics industry and its customers. Today, PVC is second only to PE in the volume of material being molded.

The repeating portion of the PVC molecule (Table 2.6) resembles the PE monomer. Both are long chains of carbon and hydrogen atoms. The difference between the two is that some of the hydrogen atoms have been replaced with chlorine atoms. As the percentage of chlorine in the molecule increases, there is an increase in density, heat, and chemical resistance.

The first rotationally moldable PVCs were liquids that adapted well to this process. These liquid PVCs came to be known as *plastisols*. Today, there are flexible, rigid, and acrylic modified liquid plastisols specifically compounded for this purpose. PVC molding materials are also available in dry powder form, with

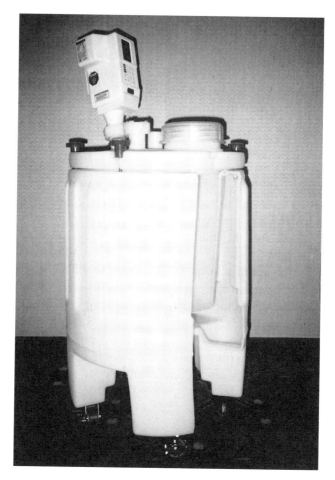

Figure 2.7 A double-walled, autoclavable, PP bioprocess and reaction vessel that replaced stainless steel (Courtesy Nalge Nunc International, Rochester, NY)

particle sizes in the range of 0.838 mm (0.033 in.), and as micropellets measuring 0.559 mm (0.022 in.). All these materials are made up of solid particles of PVC in a plasticizer. The liquid plastisols are highly plasticized, while the dry materials are only lightly plasticized.

The liquid plastisols are lower in cost. They are more difficult to handle in manual molding operations, but they allow easy automation of the metering and mold charging process for large-volume production, such as balls and doll parts.

Table 2.6 Summarizing PVC

$$\left[\begin{array}{cccc} \text{H} & \text{H} & \text{H} & \text{H} \\ | & | & | & | \\ \text{C} & \text{C} & \text{C} & \text{C} \\ | & | & | & | \\ \text{H} & \text{Cl} & \text{H} & \text{Cl} \end{array}\right]_n$$

Properties	Standard	Unit	PVC
Tensile yield	D-638	MPa	5.5 ... 16.2
		psi	800 ... 1,200
Flexural modulus	D-790	MPa	448.20 ... 2102.16
		psi	65,000 ... 305,000
Heat-deflection temperature	D-648	°C @ .455 MPa	20 ... 30
		°F @ 264 psi	68 ... 86
Izod impact (notched)	D-256	J/m	26.7 ... N.B.
		ft.lb./in.	0.5 ... N.B.
Density	D-1505	g/cm^3	1.18 ... 1.32
Mold shrinkage	D-955	%	1.0 ... 2.5

N.B. = No Break

A wide range of stiffness can be compounded. Transparent grades are available. Extremely fine details can be molded.

The dry powdered PVCs are more costly. They can be handled using common PE powder equipment. A wide range of hardnesses is available. These materials are noted for their ability to produce uniform wall thickness.

Small-volume processors purchase plastisols precompounded in barrels. Some larger volume processors purchase the individual ingredients and make their own plastisol in-house. In-house compounding results in reduced cost and the ability to make special molding materials, including highly filled and foaming formulations.

All the PVCs are known for their soft, supple feel, and their ability to be easily recompounded as rigid materials. PVCs are the best of the common rotationally molded materials for picking up fine surface details. They are relatively low in cost, but they have a high density. These materials have good chemical resistance, are nonburning, and exhibit good weatherability. These transparent materials are easily pigmented to produce bright colors or mottled effects, with good long-term color stability. Solvent bonding is a common assembly procedure. The PVCs are easily decorated and can be painted without any pretreatment. Labels and decals form a secure bond to PVC.

One unique characteristic of plastisols is that they do not develop their full rigidity until after they have been demolded. This allows the production of rigid and flexible parts with deep undercuts that would be impossible to produce with other materials.

For example, molded dolls' heads are collapsed and demolded through the neck portion of the cavity. This allows the production of aesthetically demanding parts, such as a doll's face or figurines, with no mold parting-line marks on the appearance surfaces.

This unusual property also allows small balls to be inflated to three or four times their as-molded diameter. If the balls are inflated shortly after molding and before the material has completely set, they will retain their expanded size.

Plastisol is limited by the fact that once the material has been cured during the molding process, it cannot be remolded by the rotational molding process. Cured plastisols have been successfully reprocessed by high shear rate processes, such as extrusion and injection molding. Moreover, plastisol's relatively low tensile and flexural strength, and especially its heat deflection temperature, limit its use in hot, load-bearing applications.

Plastisol parts can be produced on all the common molding machines now in use.

The liquid plastisols are dispersants of rigid PVC in liquid plasticizer. At room temperature, the particles do not dissolve. Solvation takes place at 60 to 93°C (140 to 200°F). At that temperature, the particles begin to swell and the material starts to fuse together, or gel. At a temperature of about 177°C (350°F), the material fuses into a solid mass.

It is mandatory that all the material be deposited onto the cavity before it gels and stops flowing. This requires accurate control of the heating cycle, and the speed and ratio of rotation. These special processing requirements make it very unlikely that PVC can be molded along with PE parts on the same molding machine arm.

Plastisol can be, and has been, molded in all the common molds used for rotational molding. If overheated to the point of degradation, the chlorine in the PVC combines with hydrogen in the air to create hydrochlorine acid (HCL). These small amounts of HCL can corrode ferrous molds and mold frames. Electroformed nickel and copper, cast, fabricated, and machined aluminium, and stainless steel fabricated and machined molds are not attacked by HCL.

Electroformed cavities find wide usage in the molding of plastisol. These cavities can incorporate the fine details that plastisol is capable of replicating. Electroforming is also the ideal way to produce the previously discussed hidden parting-line cavities, of the type used for dolls' heads.

The physical properties of the rotationally moldable PVCs are summarized in Table 2.6.

2.8.1   Applications

- Industrial products: flexible and rigid air ducts, machine feet, air and water filters, gaskets, tires, mannequins, and floor scrubber squeegees and bladders.
- Medical products: examination chair arms, flexible anesthesia face masks (Fig. 2.8), ear and ulcer irrigating syringe bulbs, blood pumps, respiration squeeze balloons, and anatomical teaching models.
- Transportation products: traffic cones and highway barriers, flexible bellows, dust covers and gearshift boots, arm- and headrests, and instrument panel skins.
- Consumer products: figurines and life-size statuary (Fig. 2.9), soft cushioning furniture, picture frames, artificial fruit, and toys.
- Recreational products: all types of play and "hoppity-hop" balls; soft, cuddly, and noise-making squeeze toys; lifelike doll and animal heads and body parts; toy wheels; energy-absorbing sports helmet liners; and boat and pier bumpers.

Figure 2.8 An inflatable, flexible, form-fitting, PVC anesthesia face mask (Courtesy Roto Plastics Corporation, Adrian, MI)

Figure 2.9 Highly decorated, rigid PVC, lifesize Ronald McDonald statue (Courtesy Dutchland Plastics Corp., Oostburg, WI)

2.9   Nylon

Nylon was the first high-performance engineering plastic suitable for the rotational molding process. Nylon is now the third most commonly used material in this industry.

The repeating part of the nylon molecule (Table 2.7) is characterized by the presence of nitrogen and oxygen, in addition to the usual carbons and hydrogens. The monomer's scientific name is polycaprolactom, and the polymer is a polyamide. These names do not roll off the tongue and the materials have come to be known as *nylons*. The different types of nylon are identified by a number representing the number of carbons in the repeating part of the molecule. The nylon molecule shown in Table 2.7 has six carbons and is referred to as *Nylon 6*. If there were eleven carbons in the molecule, it would be identified as *Nylon 11*.

The three polymers that are routinely rotationally molded are Nylon 6, Nylon 11, and Nylon 12. Each polymer is different, but they share many attributes. All are linear molecules that can pack tightly together. All are crystalline in nature. As a group, they are known for their strength, creep resistance, toughness, fatigue resistance, and high service temperature. Like other nylons, they resist abrasion and find use as self-lubricating bearings. They have excellent chemical resistance and can be used in contact with a wide range of reagents, including hot water and hydrocarbons such as gasoline and diesel fuel. The primary disadvantages of nylon are that they have a narrow processing window, there are no transparent grades, they are relatively high in cost, and they are hygroscopic.

All the nylons absorb moisture. At full saturation, Nylon 6 can absorb up to 9.5% of water by weight; Nylon 11 and 12 can absorb only 1.9% and 1.5% respectively [12]. The presence of moisture in the material affects its physical properties. Any water in the material turns to steam during the molding process. A just-molded nylon part will be perfectly dry. This part will exhibit high stiffness and low impact strength. The moisture in nylon acts as a plasticizer or intermolecular lubricant. As a dry part absorbs moisture from the atmosphere, it will regain its toughness and lose some of its stiffness.

The absorption of moisture also results in an increase in linear dimensions. This swelling of the material and the resulting change in dimensions must be allowed for in the sizing of the cavity of the mold. These changes in properties and size are not unique to rotational molding, and are present with all nylon molding processes. This situation is unsettling to the uninitiated, but it is well understood and allowed for by anyone familiar with nylon.

Among the three types, Nylon 6 is always the first choice, due to its lower cost. It also has the highest service temperature and overall strength, with the exception of toughness. This material's low coefficient of linear thermal expansion is useful in applications where large parts have to perform attached to metal structures with a lower rate of thermal expansion.

Nylon 11 and 12 cost more, but they are chosen over Nylon 6 because of their ease of processing, their lower density, their lower moisture absorption rate, and their dimensional stability. Nylon 11 and 12 can sometimes be molded on the same arm along with PE parts. These two nylon materials have the ability to pick up finer surface details, while producing smoother surfaces. Nylon 12 is reported to have better resistance to the newer modified gasolines, which contain a wide array of additives.

All nylons are notch-sensitive. They do not yield high values when tested by the frequently used notched Izod impact testing procedure (ASTM D-256). This

Table 2.7 Summarizing Nylon

$$\left[-N - \overset{\overset{\displaystyle H}{|}}{\underset{\underset{\displaystyle H}{|}}{C}} - \overset{\overset{\displaystyle H}{|}}{\underset{\underset{\displaystyle H}{|}}{C}} - \overset{\overset{\displaystyle H}{|}}{\underset{\underset{\displaystyle H}{|}}{C}} - \overset{\overset{\displaystyle H}{|}}{\underset{\underset{\displaystyle H}{|}}{C}} - \overset{\overset{\displaystyle H}{|}}{\underset{\underset{\displaystyle H}{|}}{C}} - \overset{\overset{\displaystyle O}{\|}}{C} - \right]_n$$ Nylon 6 Polycaprolactom

Properties	Standard	Unit	Nylon 6	Nylon 11	Nylon 12
Tensile yield	D-638	Mpa	51.7...72.4	44.8...62.0	25.0...40.0
		psi	7,500...10,500	6,500...9,000	3,620...5,800
Flexural modulus	D-790	Mpa	1378.8...2619.7	1172.0	1172.0...1240.9
		psi	200,000...380,000	170,000	170,000...180,000
Heat-deflection temperature	D-648	°C @ 0.455 Mpa	176.7	150.0	136.1...150.0
		°F @ 264 psi	131...149	113	122...125
Izod impact (notched)	D-256	J/m	48.0	74.7	32.0
		ft.lb/in.	0.9	1.4	0.6
Low-temperature impact (0.635 cm)	ARM	-40°C J	61.0*	89.5	N.A.
		-40°F ft.lb	45**	66	N.A.
Density	D-1505	g/cm³	1.130	1.030...1.040	1.010...1.030
Mold shrinkage	D-955	%	1.5...2.0	2.5	1.4...3.0
Melt index	D-1238	g/10 min	4.5...10.0	N.A.	N.A.

N.A. = Not applicable or not available
*@ -29°C
**@ -20°F

is especially true for an as-molded dry nylon part. Once these materials absorb moisture and are stabilized, Nylon 11 and 12 will have a lower stiffness and a higher elongation than Nylon 6. In the end-use situation, properly molded nylon 11 and 12 will normally be tougher than Nylon 6.

Two-component liquid nylons have been perfected for the reaction injection molding industry. These materials have an exothermic reaction and cure into a solid part at low mold temperatures. These liquid polymers are the best of the nylons for incorporating large amounts of fillers or reinforcing fibers. The liquid nylon materials are now starting to be used for rotational molding, and it can be anticipated that their usage will increase in the future.

Nylon can be molded on all the different molding machines. The primary difference in processing is that nylon is susceptible to oxidation by the oxygen in the air in the cavity during the heating portion of the cycle. This oxidation can be detected as a tan or brown color on the inner surface of a molded part. This situation can be overcome by purging the cavity with nitrogen or carbon dioxide. Failure to purge the cavity will result in a significant loss in physical properties. Some molders have reported success in molding Nylon 11 and 12 without a nitrogen purge by keeping the oven temperature to a minimum. All nylons are hygroscopic. Nylon molded with a moisture content of more than 0.2% will suffer a loss in physical properties.

In 1982, Allied introduced Nylon 6 as small pellets, measuring 1.6 mm (0.062 in.) in diameter, with a length of 2.5 mm (0.100 in.). Molding the nylon in pellet form avoids the cost of pulverizing. These small pellets are slower to absorb moisture because they have less surface area than powder. Nylon 6, 11, and 12 are also available as thirty-five mesh powder. These powdered forms of nylon can mold walls as thin as 1.5 mm (0.060 in.). The use of powdered nylon also produces smoother finishes and eliminates the surface porosity and pockmarks that are a limitation of molding pellets. Powders melt faster than pellets which, in some instances, has minimized the heating portion of the cycle. Blends of pellets and powder have been used to produce thin-walled parts of complex or unusual shapes.

All the common molds can be used to mold nylon. Nylon requires good quality molds with no undercuts, smooth surfaces and properly fitting parting lines. Nylon is a stiff material, which makes it difficult to demold parts designed with undercuts or deep textures on inside surfaces. Undercuts should be avoided. A minimum draft angle of 1.5° per side is recommended.

The crystalline nature of nylon requires slow, gentle cooling in order for the crystallinity to develop. Rapid cooling will increase warpage and reduce the material's physical properties, especially chemical resistance and stiffness. Nylon 11 and 12 are less crystalline than Nylon 6.

The physical properties of the rotationally moldable grades of nylon are summarized in Table 2.7.

2.9.1   Applications

Nylons compete primarily with PE in the marketplace. Both materials are used for many of the same applications. End users specify PE wherever possible because of its low cost and ease of processing. Nylon becomes the choice when PE isn't quite good enough. This is especially true when the application requires more temperature resistance, tensile strength, or chemical resistance in contact with oil and gasoline. The U.S. Army chose nylon instead of PE for the Hummer tactical vehicle that saw service in Operation Desert Storm (Fig. 2.10). A lift truck manufacturer tested both PE and nylon for its hydraulic oil tanks. Nylon was chosen because of its superior performance at operating temperatures up to 82°C (180°F). In each case, nylon got the job because of its improved temperature resistance, coupled with its ability to be used in contact with hydrocarbons. In other instances, nylon is chosen over PE because it is easier to paint and results in a more durable coating.

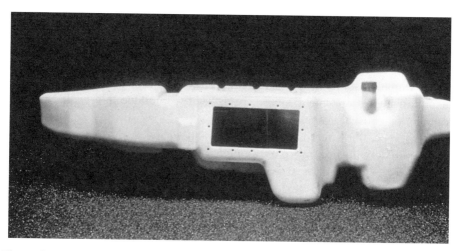

Figure 2.10 Nylon tank used on the U.S. Army's Hummer tactical vehicle (Courtesy Association of Rotational Molders)

Typical uses for nylon include fuel storage tanks, high temperature ducts, truck radiator surge tanks, large chemical shipping bottles, air horns, pressurized water treatment tanks, abrasion-resistant skids, cyclone separator housings, fuel tank level-indicating floats, aerial seeding distributors, air intake manifolds, and an Audi air filter housing.

2.10 Polycarbonate

The first rotationally moldable plastic material that provided both transparency and a high service temperature was PC. This material was first molded in 1968. Today, PC is the fourth most common material being rotationally molded.

The repeating part of the PC molecule (Table 2.8) contains benzene rings and oxygen, in addition to carbon and hydrogen atoms. These repeating units

Table 2.8 Summarizing PC

Property	Standard	Unit	PC
Tensile yield	D-638	MPa	64.1
		psi	9,300
Flexural modulus	D-790	MPa	206.8
		psi	330,000
Heat-deflection temperature	D-648	°C @ .455 MPa	131.1
		°F @ 264 psi	268
Izod impact (notched)	D-256	J/m	907.4
		ft.lb./in.	17.0
Density	D-1505	g/cm^3	1.20
Mold shrinkage	D-955	%	5 ... 7
Melt index	D-1238	g/10 min	11

combine to create molecules that do not pack tightly together. The resulting polymer is amorphous.

PC is a transparent, high-performance engineering material with high strength, and good impact and temperature resistance. With a mold shrinkage factor of only 0.5 to 0.7%, PC is more dimensionally stable than the four other commonly molded materials. The low shrinkage factor also minimizes warpage. PC has excellent outdoor weatherability, and self-extinguishing grades are available. This material can be painted without pretreatment. Solvent bonding is a common assembly technique.

PC's limitations are its relatively high cost, limited chemical resistance, and high processing temperature. These transparent materials can be colored, but care must be taken in choosing pigments that will withstand the high oven temperatures. PC is a notch-sensitive material that suffers a significant loss in impact strength if molded as a part with sharp corners. This material cannot pick up as fine surface details as some of the other commonly molded materials. PC is only available in a few grades from a limited number of suppliers.

PC can be molded on all the common molding machines now in use.

PCs are hygroscopic materials that can contain 0.35 to 0.58% moisture. Molding predried material will enhance both clarity and physical properties. Some heavily textured parts have been successfully molded without drying. Molding with wet PC results in bubbles in the wall and a yellowing of the inner surface.

All the common molds have been used with PC. The surface porosity that is sometimes present on cast aluminium and electroformed cavities traps moisture and mold release. The best results are obtained with machined or fabricated aluminium or steel molds.

PC is a strong, rigid material. Well drafted, smoothly polished cavities with no undercuts are required for efficient demolding. PC competes with PE in applications requiring low-temperature impact strength. PVC is the only other moldable material that is transparent. In applications where stiffness and temperature resistance are important, the only other choice is nylon. PC becomes the material of choice for products that can benefit from a combination of transparency, impact strength, and high service temperature.

The physical properties of PC are summarized in Table 2.8.

2.10.1 Applications

The largest single market for PC is as transparent and translucent lighting globes of all types (Fig. 2.11). PC's impact strength allowed it to take this application

Figure 2.11 Shatter-resistant PC lighting globes (Courtesy The Niland Company, El Paso, TX)

away from glass, first in locations where breakage was a problem, but now in all areas. This material's excellent outdoor weatherability is another important consideration in this application.

Other uses include pressurized beer containers and dispensers, air-cleaner housings, heating and intake ducting, space-saving airline trash containers that meet Federal Aviation Administration requirements, illuminated traffic signage, transparent food and medical containers and handling devices that require autoclaving or steam cleaning, and components requiring precise dimensions and a minimum of warpage.

2.11   Material Selection

Rotational molding is a plastic material based industry. The process imposes severe limitations on the materials that can be used. The success of a rotational molding project is more dependent on the material being molded than most of the other plastics processes.

The ideal plastic material for this process would

1. have the heat resistance and strength of steel;
2. be available in flexible or rigid forms;
3. have excellent weatherability and chemical resistance;
4. be as clear as glass, but easily colored;
5. be available in a powder or liquid form;
6. flow like water;
7. have broad processing latitude;
8. be dimensionally stable, with low shrinkage;
9. have a low density; and
10. be low in cost.

Regrettably, there is no plastic material that will meet all these requirements. The selection of the optimum material for a given application will, therefore, be a compromise between what is needed and what is available.

Considering rotational molding's dependence on the material being molded, it is mandatory that a specific plastic be chosen before the part design is finalized or mold construction is initiated.

There are many ways to select a suitable plastic material for a given application. Independent of which approach is taken, the first action has to be to develop a clear understanding of the requirements of the product in its end-use environment. One way of achieving this important objective is with the aid of a product development checklist (Table 2.9). This checklist is intended to be used for the rotational molding of plastic materials. It is a generic checklist that is not specific to any product. The functional requirements of large, stationary PE storage tanks are very different from those of a small, flexible PVC automobile armrest. The checklist is intended only as a guide. The items that don't apply to a specific product are to be crossed off and replaced by the unique requirements of the product being considered.

The checklist is intended for, but is not limited to, use by a new product design and development engineer working for an OEM. In this environment, the ideal scenario would be for the design engineer and the company's ultimate customer(s) to fill out the checklist together. The global marketing and mass merchandising approaches now being practiced often make it impractical for a design engineer to meet face-to-face with customers. The next best approach is a meeting with the company's sales or marketing personnel, who are in direct contact with the customer and are familiar with the functional requirements of the product.

It will be an unusual situation when all the questions posed in the checklist can be answered. In most cases, additional research will be required in order to

Table 2.9 Product Development Checklist

Part or Product Name: _____

Description of Application: _____

_____

Prepared By: _____ Date: _____

Physical Requirements:

    Size: Length: _____Width: _____ Height: _____

    Volumetric Capacity (liters, kilograms, etc.): _____

    Maximum (Minimum) Weight: _____

    Density (Maximum or Minimum): _____

    Must fit and function with: _____

        Critical Dimensions: _____

    Other: _____

Mechanical Requirements:

    Tensile loading:

        Load: _____ Type: _____

        Duration: _____

    Flexural loading:

        Load: _____ Type: _____

        Duration: _____

    Stiffness (Flexural Modulus): _____

    Compressive loading: _____

    Creep—maximum allowable: _____

    Deflection—maximum allowable: _____

    Impact strength:

        At room temperature: _____

        High temperature: _____

        Low temperature: _____

    Shear strength: _____

    Hardness: _____

    Abrasion resistance: _____

    Other: _____

Environmental Requirements:

    Operating temperature:

        Maximum: _____

        Minimum: _____

        Duration: _____

(*continued*)

Thermal expansion limits: _____

Flammability requirements: _____

Outdoor exposure _____

Chemical resistance

    Continuous contact with _____

    Intermittent contact with _____

    Occasional contact with _____

Painting system:

    Solvent attack: _____

    Oven temperature: _____

Vapor permeability: _____

Other: _____

Electrical Requirements:

Volume resistivity: _____

Surface resistivity: _____

Dielectric constant: _____

Dissipation factor: _____

Dielectric loss: _____

Arc resistance: _____

Electrically conductive: _____

Transparent to microwaves: _____

EMI/RFI shielding: _____

Other: _____

General Requirements:

Durable (life-expectancy): _____

Disposable or reusable: _____

Quantity required: _____

Target factory cost: _____

Patent restrictions: _____

Product for export—where: _____

Recyclability: _____

Protective packaging: _____

Anticipated misuse, safety hazards, and warnings: _____

Material now used: _____

Process now used: _____

Other: _____

Appearance Requirements:

    Industrial design required: _____

        Aesthetics: _____

        Human engineering: _____

    Match existing product line: _____

    Transparent: _____

    Color—two color: _____

    Decorating:

        Surface finish (SPI # _____) Other: _____

        Texture: _____

          Texture depth: _____ Pattern No.: _____

        Flame polishing: _____

        Painting: _____

        Waxing: _____

        Metalizing: _____

        Hot stamping: _____

        Pad printing: _____

        Screen printing: _____

        Other: _____

    Labeling requirements

        Engraved in mold: _____

        Stick-on: _____

        Molded-in: _____

        Decal: _____

        Stencil: _____

        Other: _____

Secondary Requirements:

    Machining required: _____

        Special fixtures required: _____

        Machining line location: _____

    Vent-tube hole location: _____

    Parting line location: _____

        Trimming required: _____

    To be assembled to: _____

    Assembly to be:

        Permanent: _____

        Serviceable: _____

(*continued*)

Leakproof: _____

Disassembly for recycling: _____

Type of assembly required:

Mechanical—screws or inserts: _____

Mechanical—press or snap-fit: _____

Adhesive: _____

Solvent: _____

Heat sealing: _____

Electromagnetic bonding: _____

Spin welding: _____

Molded threads or fasteners: _____

Other: _____

Flexibility in design of mating part: _____

Special quality tests: _____

Regulatory Agency Approval Required:

American National Standards Institute (ANSI): _____

American Society of Testing and Materials (ASTM): _____

Department of Transportation (DOT): _____

Federal Communication Commission (FCC): _____

Food and Drug Administration (FDA): _____

Military Specifications: _____

National Electrical Manufacturers Association (NEMA): _____

National Sanitation Foundation (NSF): _____

Occupational Safety and Health Administration (OSHA): _____

Society of Automotive Engineers (SAE): _____

State or local building codes: _____

U.S. Pharmacopeia (USP): _____

Underwriters Laboratories (UL): _____

Flammability rating (UL–94): _____

Temperature index (UL–746): _____

Other: _____

For export: _____

Canadian Standards Association (CSA): _____

International Standards Organization (ISO): _____

Others: _____

answer these questions. In these times of simultaneous engineering and early market introduction, there is a natural hesitancy to take the time and expend the effort required to complete the checklist. The benefits to be derived from this investment in time are definitely worth the effort. One of the surest ways to realize an early market introduction is to take the time to do the project right the first time. One of the best ways to avoid mistakes is to thoroughly understand what is required before any work is begun on the project.

A carefully completed product development checklist becomes a product specification that an engineer can use during the design phase of a project. The same information can be used in selecting the optimum combination of plastic material and processing techniques. The same data will also be helpful in determining what kind of mold will be required. The checklist becomes a permanent record of what was agreed to at the onset of the project. This eliminates a lot of misunderstanding and disappointment that frequently accompany the sampling of a new mold. If some important detail was left out of the original product specification, no one has a right to expect that detail to be present in the finished part.

Once the new product design has been roughed out and approved, the engineer then selects a process that is capable of producing the required part. The next step is to choose a plastic material that is suitable for the product and the process. There were 24,000 different plastic materials being offered for sale in North America in 1998. Design engineers are accustomed to choosing a plastic material from this large list. The engineer considering rotational molding for the first time will be surprised to discover that only a few of these plastic materials are suitable for this process.

One way of starting the critical material selection process is to run copies of the product development checklist. Those details that are material-related can then be marked and studied. With this information in mind, the design engineer is ready to begin the material selection process. One of the oversights that leads to mistakes in the choice of a material is the failure to thoroughly define what is required. All too often, the design engineer chooses a material based on two or three key properties, without taking all of the requirements into consideration. Many plastic products have failed because someone forgot about chemical resistance while concentrating on cost, tensile strength, and temperature resistance. The checklist is helpful in preventing the design engineers from forgetting about important but obscure requirements, such as the Department of Transportation's regulations.

If rotational molding is to be the manufacturing process, then every effort must be made to find a usable plastic among the commonly molded materials listed in Table 2.1. The most important physical properties of the five families of commonly molded plastics are shown in Table 2.10. The data are the maximum

Table 2.10A Common Material Properties; Metric Units Spreadsheet

Property	ASTM Standard	Low Density Polyethylene	Linear Polyethylene	Medium Density Polyethylene	High Density Polyethylene	Cross-Linked Polyethylene	Poly-propylene	Ethyl Vinyl Acetate Copolymer	Polyvinyl Chloride Plastisol	Polyvinyl Chloride Dry Powder	Nylon 6	Nylon 11	Nylon 12	Polycarbonate
Tensile Yield (MPa)	D-638	7.9 / 11.0	13.1 / 20.7	14.5 / 18.5	20.7 / 27.6	17.9 / 24.1	19.2 / 27.6	9.7 / 11.7	5.5 / 8.3	9.0 / 16.2	51.7 / 72.4	44.8 / 62.0	25.0 / 40.0	64.1
Flexural Modulus (MPa)	D-790	138.6 / 365.4	427.4 / 854.9	689.4 / 896.2	930.7 / 1447.7	482.6 / 792.8	1206.5 / 1365.0	43.7 / 95.8	N.A.	N.A.	1378.8 / 2619.7	1172.0	1172.0 / 1240.9	206.8
Heat Deflection Temperature (°C @ 0.455 MPa)	D-648	23.0	87.8 / 97.8	51.7 / 62.8	67.2 / 71.7	65.0 / 66.1	68.9 / 97.8	37.2	N.A.	N.A.	176.7	150.0	136.1 / 150.0	135.0
Heat Deflection Temperature (°C @ 1.82 MPa)	D-648	−12.0	42.2 / 50.0	38.9	41.7 / 43.9	46.1 / 66.7	45.0 / 56.1	22.8	N.A.	N.A.	55.0 / 65.0	45.0	50.0 / 51.7	131.1
Low Temperature Impact (0.635 cm) (J)	ARM −40°C	81.3	33.9 / 81.3	38.0 / 271.2	135.6 / 183.0	94.9 / 135.6+	27.1 / 61.0	264.4 / 271.2	N.A.	N.A.	61.0*	89.5	N.A.	N.A.
IZOD Impact (notched) (J/m)	D-256	389.7 / N.B.	774.0 / N.B.	133.4 / 160.1	245.5 / 389.7	53.4 / 1067.6	26.7 / 117.4	N.B.	533.8	N.A.	48.0	74.7	32.0	907.4
ESCR (100% IGEPOL) (Hours)	D-1693	1.0 / 1.5	8.0 / 1000+	1000+	5.3 / 31.0	1000+	1000+	4.0	N.A.	N.A.	N.A.	N.A.	N.A.	N.A.
Hardness A, D, or R	D-2240	46D / 52D	N.A.	52D / 58D	60D / 66D	N.A.		N.A.	55A / 98A	73A / 95A	116R	78R / 113R	85R / 110R	118R / 123R
Density (g/cm^3)	D-1505	0.910 / 0.925	0.926 / 0.940	0.926 / 0.940	0.941 / 0.959	0.933 / 0.945	0.900	0.927 / 0.941	1.180 / 1.320	1.400	1.130	1.030 / 1.040	1.010 / 1.030	1.200
Mass (cm^3/g average)		1.091	1.073	1.073	1.053	1.066	1.112	1.072	0.801	0.715	0.886	0.967	0.981	0.834
Melt Index (g/10 min)	D-1238 - E	4.5 / 22.0	2.5 / 14.3	5.0 / 8.0	5.0 / 8.0	N.A.	5.0 / 20.0	11.0 / 23.0	N.A.	2.4 / 56.6	4.5 / 10.0	N.A.	N.A.	11.0
Mold Shrinkage (%)	D-955	1.5 / 2.0	2.0 / 2.2	N.A.	1.5 / 3.0	2.0 / 3.0	1.0 / 2.5	1.5 / 3.5	1.0 / 2.5	N.A.	1.5 / 2.0	2.5	1.4 / 3.0	0.5 / 0.7
Thermal Conductivity (cal/s/cm/°C $\times 10^{-4}$)		8.0	N.A.	8.0	11.0 / 12.0	N.A.	2.8 / 4.0	3.0 / 4.0	3.0 / 4.0	3.0 / 4.0	5.8	8.0	5.2 / 7.3	4.7
FDA Grades Available		YES	YES	YES	YES	NO	YES	NO	YES	N.A.	NO	N.A.	N.A.	YES
Reprocessable		YES	YES	YES	YES	YES	N.A.	N.A.	NO	NO	N.A.	N.A.	N.A.	YES

N.A. = Not Applicable or Not Available   N.B. = No Break   *@ −29°C

Table 2.10B Common Material Properties; English Units Spreadsheet

Property	ASTM Standard	Low Density Poly-ethylene	Linear Poly-ethylene	Medium Density Poly-ethylene	High Density Poly-ethylene	Cross-Linked Poly-ethylene	Poly-propylene	Ethyl Vinyl Acetate Copolymer	Polyvinyl Chloride Plastisol	Polyvinyl Chloride Dry Powder	Nylon 6	Nylon 11	Nylon 12	Polycarbonate
Tensile Yield	D-638 psi	1,140 / 1,600	1,900 / 3,000	2,100 / 2,690	3,000 / 4,000	2,600 / 3,500	2,790 / 4,000	1,400 / 1,700	800 / 1,200	1,300 / 2,350	7,500 / 10,500	6,500 / 9,000	3,620 / 5,800	9,300
Flexural Modulus	D-790 psi	20,100 / 53,000	62,000 / 124,000	100,000 / 130,000	135,000 / 210,000	70,000 / 115,000	175,000 / 198,000	6,340 / 13,900	N.A.	N.A.	200,000 / 380,000	170,000	170,000 / 180,000	30,000
Heat Deflection Temperature	D-648 °F @ 66 psi	73.4	190 / 208	125 / 145	153 / 161	149 / 151	156 / 208	99	N.A.	N.A.	350	302	277 / 302	275
Heat Deflection Temperature	D-648 °F @ 264 psi	10.4	108 / 122	102	107 / 111	115 / 152	113 / 133	73	N.A.	N.A.	131 / 149	113	122 / 125	268
Low Temperature	ARM −40°F	N.A.	25	28	100	70	20	195	N.A.	N.A.	45*	66	N.A.	N.A.
Impact (¼")	ft.lbs.	60	60	200	135	100+	45	200	N.A.	N.A.	0.9	1.4	0.6	17
IZOD Impact (notched)	D-256 ft.lb/in.	7.3 / N.B.	14.5 / N.B.	2.5 / 3.0	4.6 / 7.3	1.0 / 20.0	0.5 / 2.2	N.B.	10 / N.B.	N.A.	N.A.	N.A.	N.A.	N.A.
ESCR (100% IGEPOL)	D-1693 Hours	1.0 / 1000+	8.0 / 1000+	1000+	5.3 / 31	1000+	1000+	4.0	N.A.	N.A.	N.A.	N.A.	N.A.	N.A.
Hardness	D-2240 A, D, or R	46D / 52D	N.A.	52D / 58D	60D / 66D	N.A.	N.A.	N.A.	55A / 98A	73A / 95A	116R	78R / 113R	85R / 110R	118R / 123R
Density	D-1505 gm/cm³	0.910 / 0.925	0.926 / 0.940	0.926 / 0.940	0.941 / 0.959	0.933 / 0.945	0.90	0.927 / 0.941	1.18 / 1.32	1.40	1.13	1.03 / 1.04	1.01 / 1.03	1.20
Mass	In³/lb ave.	30.196	29.69	29.69	29.16	29.50	30.78	29.66	22.16	19.79	24.52	26.77	27.16	23.09
Melt Index	D-1238 - E	4.5 / 22.0	2.5 / 14.3	5.0 / 8.0	5.0 / 8.0	N.A.	5.0 / 20.0	11.0 / 23.0	N.A.	2.38 / 56.58	4.5 / 10.0 -	N.A.	N.A.	11.0
Mold Shrinkage	D-955 %	1.5 / 2.0	2.0 / 2.2	N.A.	1.5 / 3.0	2.0 / 3.0	1.0 / 2.5	1.5 / 3.5	1.0 / 2.5	N.A.	1.5 / 2.0	2.5	1.4 / 3.0	0.5 / 0.7
Thermal Conductivity	cal/cm²/sec Per °C cm×10⁻⁴	8.0	N.A.	8.0	11.0 / 12.0	N.A.	2.8 / 4.0	N.A.	3.0 / 4.0	3.0 / 4.0	5.8	8.0	5.2 / 7.3	4.7
FDA Grades Available		YES	YES	YES	YES	NO	YES	NO	YES	N.A.	NO	N.A.	N.A.	YES
Reprocessable		YES	YES	YES	YES	YES	N.A.	N.A.	NO	NO	N.A.	N.A.	N.A.	YES

N.A. = Not Applicable or Not Available   N.B. = No Break   *@ −20°F and minimum values for the different grades of materials in each family category.

The best plastic for any application is the lowest cost material that will satisfy the product's requirements. The search should, therefore, start with an analysis of LLDPE. If this material is lacking in some way, the next more costly material becomes the prime candidate. In some instances, there will be an ideal material for an application. In most cases, there will not be one single plastic material that will satisfy all the product's requirements. Most material selections involve compromise. For example, the application might need nylon's temperature resistance, but cost constraints favor LLDPE. This brings up the question of what was asked for and wanted, versus what is actually needed.

If a usable material cannot be found among the commonly molded plastic materials, the design engineer should question the troublesome requirements in the design checklist. For example, is a UL-94 VO flammability rating mandatory or just desirable? It might be nice to have the product in a transparent material, but would a colored part be good enough?

If the requirements of the product cannot adapt to the limitations of the common materials, then the design engineer has no choice and must then turn to the other rotationally molded plastic materials listed in Table 2.2. While searching through the other materials, it must be remembered that not all these materials are commercially available. There is also a limited number of custom molders who know how to process these other materials.

If a usable material cannot be found among the other plastics, the design engineer's last resort is a specially compounded or modified material, as described in Section 2.4.

A plastic material has many functions to perform, but these requirements can be separated into two categories. First, the material must satisfy the functional requirements of the product in the hands of the end user. Second, the material must meet the requirements imposed by the manufacturing process being used. Both requirements have to be satisfied, but functional requirements must be given priority. If the functional requirements of the end user are not satisfied, a customer will not purchase the product. If the product does not sell, there is no need for efficient manufacturing.

If the product specification cannot change and no usable material can be found among the common, other, or special materials, the project may not be suitable for the rotational molding process. In that case, the design engineer should consider the other thermoplastic hollow part processes of extrusion blow molding and twin-sheet thermoforming. If neither of these processes is suitable, the last alternative would be an assembly of two or more parts produced by one of the many other plastics manufacturing processes. That would be a loss to the rotational molding business. The industry does not like losing a project, but it is better than proceeding with a rotationally moldable material that does not satisfy the functional requirements of the product.

Even if an ideal material can be found among the commonly molded plastics, there are still many questions that must be answered. Table 2.10 is useful as a selection screening tool, but it is not specific to any one material. The manufacturer and the specific type of LLDPE still have to be chosen.

Other technical concerns, such as the linear coefficient of thermal expansion, regulatory agency compliance, weatherability, and chemical compatibility, must still be resolved. The best source for this type of detailed information is the plastic material manufacturers. Manufacturers normally know important details, such as what percentage of the original tensile strength will be lost due to five years of outdoor exposure in Arizona. Manufacturers maintain long lists of chemical reagents that can and cannot be used with their materials.

Plastic material manufacturers accumulate more experience with their materials than anyone else in the industry. They have lived through their materials' successes and failures in many applications. Material manufacturers can certainly provide input on the suitability of their material for a given application. In most cases, they can critique part designs, confirm process selection, and recommend mold types and processing conditions. The benefit of their experience is available for the asking.

There are 24,000 commercially available plastic materials, and all of them must be considered in selecting the optimum plastic for a given application. The first secret to selecting the best plastic material is to start with a clear understanding of what is required. It is more productive to spend time defining what is required than to waste time searching through all the available technical literature.

The second secret to simplifying the selection process is to limit the total number of materials being considered. In the case of rotational molding, it is possible to eliminate a lot of materials by concentrating on only those materials that are suitable for this process. Simplification through elimination can go even further. If only the commonly molded materials are being considered, and if transparency is required, the choice is immediately reduced to PVC or PC. Service temperature is another quick screening tool. Any product that has to function at 100°C (212°F) immediately eliminates PE, PP, and PVC from any further consideration.

Custom rotational molders are frequently called upon to suggest a specific plastic material for an application. Processors are certainly knowledgeable regarding what can and cannot be done with the commonly rotationally molded plastic materials. A processor may or may not know very much about the functional requirements of the product in its end-use environment. There is nothing wrong with asking a processor to suggest a plastic material. It must, however, be remembered that the final decision for the suitability of a plastic material for its intended purpose resides with the OEM, who introduces that product into the stream of commerce.

Independent of which plastic is finally chosen, that material, molded into its final form, must be thoroughly tested to confirm that a suitable material has been selected.

2.11.1   Material Information Sources

The purchasers and suppliers of rotationally molded parts share a need for a reliable source of plastic material information. There is so much data available from so many providers that it is difficult to know which source should be used. This problem is compounded by the fact that the data keep changing as standard materials are improved and new materials are introduced. Some of the sources develop their data using different test methods, making it difficult to directly compare several materials.

The best overall source for plastic material related chemical information is the material manufacturers and their published literature. A listing of the different materials being supplied by North American plastic material manufacturers can be found in handbooks, the *International Plastics Selector*, and the *Modern Plastics Encyclopedia*.

The disadvantage of these printed data is that they quickly become obsolete. The user of last year's brochure has no way of knowing whether or not the material has changed or is still available. An increasing number of the major manufacturers now post their currently available offering of materials, along with varying amounts of technical data, on the Internet. It is assumed that this information is current.

The manufacturers of plastic materials are the best source for physical properties, performance, processing, and applications information. For under-standable reasons, manufacturers limit the information published to just the materials that they produce. OEMs and molders have a need to know about all the plastic materials that are available to them and not just those being sold by a single manufacturer.

Injection molding and compression molding are the most often used processes for preparing laboratory test specimens. The low-pressure rotational molding process does not densify the plastic material to the same degree as these two high-pressure processes. The most reliable data are obtained by testing rotationally molded test specimens.

An excellent source that contains only rotational molding grades of materials information is the *Listing of Resin Properties*. With the exception of density and melt index, all these technical data are obtained using rotationally molded test specimens. The *Listing*, which is updated every two or three years, is available from ARM [13].

The *International Plastics Selector* is the most comprehensive printed source for plastic material information. This is not a rotational molding specific source, but it does contain the most complete coverage of all the available plastic materials. The *Selector* is updated annually. The current issue contains more than 24,000 plastic materials being offered for sale in the United States. The *Selector* is available from D.A.T.A. Business Publishing [14].

The product design community's most frequently used reference for plastic material information is the *Modern Plastics Encyclopedia*. This is not a rotational molding specific listing. The *Encyclopedia* is published once a year as a part of a subscription to *Modern Plastics* magazine. Subscription information is available from *Modern Plastics* [15].

There are many computer software programs and online databases that cover the physical properties of plastic materials. All these programs are only as good as the data received from the material manufacturers who supply the information. None of these programs are rotational molding specific. The most comprehensive and user-friendly material computer database is available from IDES, Inc. [16]. These data are updated quarterly.

In the United States, the physical properties of plastic materials are determined by a rigidly controlled series of test procedures that are issued and maintained by the American Society of Testing & Materials (ASTM). Over the years, these testing procedures have been modified and supplemented to accommodate special materials. Many of the testing protocols now contain subsections that allow test specimens to be prepared by different techniques. The rate of loading can be varied. For example, Izod impact testing can be performed using notched or unnotched specimens, with different sizes of radii at the bottom of the notch. Not all manufacturers test their materials with exactly the same test procedures. This lack of uniformity can make it difficult and sometimes impossible to directly compare several different materials being considered for an application. A similar situation has developed in other major plastic-producing countries, such as with Germany's DIN standards, Great Britain's BS, France's AFNOR, and Japan's JIS. There is little correlation between the test protocols promulgated by these fiercely independent agencies.

The rise of the global market and emergence of multinational OEMs has resulted in a need for a standardized method of testing and reporting plastics-related data. A uniform worldwide system would greatly simplify the free flow of commerce between trading countries. The Geneva-based International Organization for Standards (ISO) has now addressed this issue. Most of Europe and ninety additional industrialized countries in the Pacific, Asia, and Africa have now adopted the ISO standard. Over 125,000 copies have been issued in Europe alone. The United States is the only large, industrialized country that has shown a reluctance to adopt these ISO standards.

Many U.S. material manufacturing companies have gone ahead on their own and have adopted the ISO standards. This is especially true for those suppliers who sell their materials to the multinational OEMs who market their products worldwide.

The basic objective of the ISO is to allow the direct comparison of plastic materials produced anywhere in the world. This is a capability that the plastic product design community has wanted for a long time. This objective is being achieved by eliminating all or most of the variations that crept into the different systems in the past. The new standards dictate the size and proportions of the test specimens. The testing and reporting protocols have been standardized, and arbitrary variations in these procedures are no longer acceptable. All the various data-acquisition and information-reporting procedures have been combined and are referred to as the Computer-Aided Material Preselection by Uniform Standards (CAMPUS). Plastic material data collected and reported according to the CAMPUS procedures are available, free of charge, to qualified customers from the individual material manufacturers who have adopted this system. These user-friendly programs use the Windows format. Users have the option of reviewing the data in metric or English units.

3   Design Considerations

There are hundreds of interrelated details that combine to produce a rotationally molded product or part. All these details can be divided into four broad categories or elements, which are

1. the design of the part;
2. the choice of the plastic material;
3. the design and construction of the mold; and
4. the actual molding of the part.

All four of these elements must be handled correctly in order to produce the optimum part.

All four elements are of equal importance; however, everything starts with and is influenced by the design of the part. Without a part design, there is no need for a material, a mold, or a molding machine. All the different plastic materials, molds and processing techniques have their own advantages and disadvantages. A successful part design is the result of the design engineer's awareness of these capabilities and limitations, coupled with meticulous attention to design details.

During the rotational molding process, a hot mold is biaxially rotated through a puddle of liquid or powdered plastic material to coat the cavity and form a part. The process is at its best producing hollow shapes with contours that blend smoothly into each other. A round ball is the ideal shape for rotational molding. The process is also capable of producing complex shapes (Fig. 1.4), in those cases where the design engineer understands and takes into account the requirements of, and designs for, the process.

3.1 Product Design

The design of a plastic product to be produced by the rotational molding process is no different than the design of a product to be made in a different material by some other process. Before design engineers can design anything, they have to know what the product is and what it has to do. One proven method of gaining an understanding of a new product's requirements is through the use of the product development checklist (Table 2.9). The use of this checklist and the benefits it provides are described in Section 2.11.

At the beginning of the product design process, the designer should run a copy of the checklist and study it carefully. All the entries relating to the design or structure of the product should be marked and carefully analyzed. If there are unanswered design-related questions in the checklist, that information must be obtained before proceeding. Some checklist entries may be in conflict with others. For example, the checklist may specify a transparent container with the lowest possible cost. The lowest-cost material is PE, but it is not transparent. Which of these two specifications is more important? These questions must be answered before proceeding. To do otherwise could result in having to redo the design work. The single most important thing that a design engineer does in a new product development project is define what is required. Many new products have failed or not achieved their full potential because, in the excitement created by a new product opportunity, no one took the time to thoroughly define what was wanted.

Once a designer has a clear understanding of what is required, the actual design work can begin. This part of the work is the conceptual design phase. This is the exciting and just plain fun part of the process. It is also one of the more important things that design engineers do. The designer's company will rely on and proceed with one of the concepts that the designer creates during this phase of the project. The designer's reputation and the company's financial well-being are put to the test with every new product. This work must be done as carefully and as thoroughly as possible.

The conceptual design phase starts with the designer studying the checklist and thinking about the product. This is a mental exercise, where the designer draws on accumulated knowledge and past experiences. While thinking about the product, it may become obvious that the functional requirements could be met with a two-piece injection-molded assembly. One mold and the assembly process could be eliminated by blow molding the product as one piece. The small number of parts required might encourage the designer to remember that rotational molding normally has a lower mold cost than blow molding and it is a better process for small quantities.

The designer will contemplate and compare the advantages of injection molding, which can use many plastic materials and provide molded features on both sides of the wall, to those of rotational molding, which, on the other hand, is a better process for large, hollow parts that have to withstand high impact loads. The challenge might then become figuring out a way to eliminate the features on the inside of the product.

Designers have the ability to imagine things before they exist. While contemplating the advantages and disadvantages of various process and material combinations, as well as many other things, mental images begin to take form. After a while, one or two of those mental pictures recur more often than others. At this point, the designer begins to doodle or sketch these images. What the designer is actually doing is experimenting with different structures in sketch form, while searching for the optimum structure.

Early market introduction is an important goal, but the creation of a new product concept is more important. The new product can only be as good as the original concept design. Many new products fail because a company proceeds with the first concept that appears to be acceptable, without waiting for the best concept, which evolves later, after the designer has had a longer time to think about the product. The creative process takes time, and it is unwise to rush this critical phase of a project.

If the checklist specified a refuse container, a good designer would immediately demand to know what kind of refuse container. Industrial or residential? What size? Will it be used indoors or out of doors? What is the cost range? How many will be required? In this case, the checklist might clarify that this container is intended for the typical homeowner to conveniently store waste and transport it to the street for pickup. It is normally located out of doors, but may be used inside an apartment or garage. Indoor use requires an odor-confining lid or cover. Considering an average family of four, it has to be large, as the pickup service charges by the number of containers handled. Considering the weight of a full container, it will have to have handles and maybe wheels. The volume to be produced is only 5,000 the first year, and 10,000 the second and third, with an ongoing need for a few thousand per year. Cost is important, but not critical. These containers will be provided by the pickup company and will not be purchased by the homeowner. A long, trouble-free service life takes priority over part cost.

As the designer thinks about this product, images of all the refuse containers observed in the past will be mentally reviewed. Some of those containers were larger than others. Some were tall, and others were short. In cross-section, they were both round and square. Some had wheels and handles. They were made out of welded steel, molded plastic, and combinations of the two.

A creative product designer is not limited by what has been done in the past. A study of existing products is, however, a good way to start a new project. A successful product has already proven its acceptability in the actual in-use environment. Actual use by the customer is always a better evaluation of a product than any pre-introduction laboratory testing program.

If the designer specifies steel for the container, there will be no concern about strength or flammability. On a volume basis, steel costs less than plastic. Steel has a long history of successful use in this application. On the negative side, steel is heavy. Welding is slower and more labor-intensive than molding. The steel will require painting and repainting.

A plastic refuse container would be lighter in weight and quieter in use. Color, labeling, handles, and other features could be molded in. A plastic container would not rust.

Plastic is not as strong as steel, but the designer will recall having seen a lot of plastic refuse containers being used. It would, therefore, be safe to assume that plastic is strong enough.

Fabricating a welded steel container would require only a minimum initial investment in tooling. Molded plastic would require a mold, but the manufacturing cost would be lower. Amortizing the mold cost over the first three years' volume of 25,000 containers would easily justify the initial investment in a mold.

If the designer chooses to pursue molded plastic, the next question then becomes which process is the best. An astute designer would already know that refuse containers have been successfully produced by rotational molding, injection molding, thermoforming, blow molding, and injection molded structural foam.

Injection molding and structural foam, followed by blow molding, would require the highest initial investment in molds, but their part cost would be low. All three processes would be ideal for refuse containers, but not for the limited quantities required.

Injection molding and structural foam would require separate molds and molding procedures for the body and lid of a refuse container. The molded part would, however, be ready to assemble with no secondary operations. Blow- and rotationally molded, and maybe thermoformed, containers could be produced in one molding operation. All three processes would produce containers that require secondary machining operations before assembly.

The need for only a few thousand containers after the third year would favor those processes that can be quickly set up and put into production. Blow molding can require hours to set up and a change from one plastic material or color to another. The required depth-of-draw ratio is challenging for thermoforming. Rotational molding appears to be the best process for further consideration.

If rotational molding is to be the process, what plastic material will be specified? Material selection is discussed in Section 2.11. In general, the ideal material for any product is the lowest cost material that will do what is required. PE is the lowest cost plastic material that is readily rotationally molded. An observant plastics product design engineer should already know, or be able to determine, that PE is currently being used for refuse containers.

All things considered, it is safe to proceed on the assumption that the new refuse container can be made by the rotational molding of PE.

Once the design engineer has decided on a specific process and material, those decisions will influence how the product is designed. An injection molded container will not be designed the same way as a part to be rotationally molded. As the designer begins to sketch various design concepts, the structures that evolve will be suitable for the chosen material and process that is being considered. That is one of the reasons the design engineer should select a material and process before actually designing the part.

After some experimental doodling, the designer will choose one or more concepts for further consideration. The resulting sketch might look like the refuse container shown in Fig. 3.1. In this sketch, the designer has chosen a tall, rectangular shape. The container is narrow in cross-section to conserve storage space and to accommodate a standard doorway. The required volume has been achieved by increasing the height. Wheels and two large handles have been provided, to help in moving a full container.

Even at this early stage, the designer's thinking is well advanced. The basic elements of strong handles and the lid-hinging brackets are beginning to evolve. A molded-in pickup truck hook recess has been included. The molding parting line has been established and molding draft angles are evident. Large radii are being considered for the bottom of the container, which will have to withstand high impact loads.

Before proceeding any further with the project, it is desirable to review the concept with an industrial designer. Industrial design is different from engineering design. In recent years, the industrial design function has expanded to encompass many aspects of product design. Two of those functions that are important to the success of this refuse container are appearance design and human engineering.

Human engineering has to do with proportioning a product so that it fits or can be conveniently used by human beings. Is the height of the container suitable for adults and children? What about the size and location of the handles? Should there be a handle on the lid? What size wheels will be required

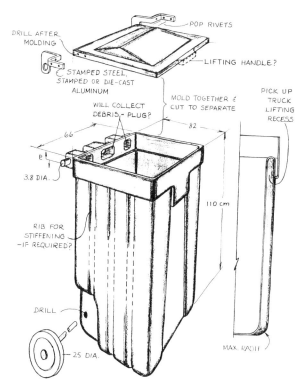

Figure 3.1 Concept sketch of a one-piece rotationally molded refuse container

for pulling or pushing the container across the lawn, or through snow or over a curb?

The appearance of an industrial product may not seem to be an important consideration. That is an incorrect assumption. Raymond Loewy, one of the founders of the North American industrial design profession, correctly observed that

"Between two products, equal in price, function, and quality, the better-looking one will outsell the other."

The carpet-cleaning machines shown in Fig. 3.2 show the beneficial influence of industrial design. Appearance is more critical with consumer products than industrial products, but it is important to both. The homeowners will feel better about using a nice-looking refuse container than an unattractive, rusted, welded, square steel box. The appearance of a product is dictated by its shape, proportions, color, and surface finish, and the way light strikes and

Figure 3.2 The housings of these carpet-cleaning machines have been designed to eliminate the stark industrial appearance of competitive products

reflects off it. It is the total image created by the product that is important. At a glance, this total image must convey the message that the product will do what it is intended to do.

Industrial designers receive special training in appearance design, human engineering, and many other subjects. The early input of a qualified industrial designer can greatly increase the customer acceptance of a product. It is beyond the scope of this book to do justice to the industrial design function. Industrial designers are, however, another valuable resource that design engineers are not using to its full advantage.

At this phase of the project, it is important to compare the product that is evolving with the design checklist in order to make certain that it is still within the original product specifications. Compromise decisions made along the way may well have eliminated some important functional features. Studies by the Institute of Competitive Design have indicated that the decisions made up to this point will determine 75% of the product's cost. The design sketches should be reviewed one last time to determine whether or not there are any changes that could be made to reduce the product's cost. In the next phase of the project, other people will become involved. Once they approve the design concept, it becomes locked in. Future changes will be more difficult to make, as they will require the approval of other interested people.

The concept sketches, or refined versions of them, will then be reviewed with the company's marketing department and the ultimate customer. If the customer is not satisfied with the structure shown in the sketch, this design concept will be rejected. The designer should then ask why the concept is unacceptable. What has to be changed to make it acceptable? The answers to these questions become valuable information that should be added to the product development checklist.

An equally likely scenario is that the concept is acceptable as sketched, or with only slight modification. If this is the case, the design engineer can proceed to develop the product.

At this phase of the project, there is a list of requirements in the form of the checklist and a sketch showing the approximate size and shape of the product. The plastic material and molding method have been chosen. If the design engineer has little or no experience with rotational molding, it is highly desirable at this point to review the project with a custom molder. A molder may not have had prior experience with, and is probably not an expert in, refuse containers, but he will be an expert in rotational molding. Custom molders are a valuable plastic product development resource and are all too often overlooked by design engineers. Any successful rotational molder will know enough about design, plastic materials, tooling, and molding to be able to comment on the manufacturability of a product. A molder can provide valuable input on finalizing the material selection and deciding what kind of mold to use, and in fine-tuning the design for efficient molding.

The sketch, with its approximate overall size and estimated wall thickness, can also be used to secure tentative tooling and molded part cost estimates. These preliminary estimates will indicate whether the product is within an acceptable price range.

Up to this point in the project, all the work that has been done can be described as product design. The next step will be to reduce the concept sketch to a detailed piece-part drawing, or computer-aided design (CAD) database.

As the project progresses from the product design phase to the part design phase, the designer's emphasis will change. All through the product design phase, the designer was working in the creative realm of searching for a structure, process, and material combination that would satisfy the product's functional requirements within an acceptable cost. This free-thinking part of the project is basically undisciplined and there are few rules to guide or restrict the designer's thinking. The only important rule is that the resulting product concept must be acceptable and, it is hoped, better than competitive products.

Piece-part design, on the other hand, is highly disciplined. By trial, error, and research, the industry has evolved a list of part design guidelines. By following these guidelines, it is possible for both experienced and novice designers to produce an acceptable part design.

3.2   Design for the Process

All plastics processes have their own special capabilities and limitations. The nature of the rotational molding process, where a hot cavity passes through a pool of cooler plastic to coat the cavity, is unique. This cavity-coating process imposes limitations on what can and cannot be produced. This process is at its best producing hollow shapes with smoothly blended contours. For example, the solid reinforcing rib (Fig. 3.3A) can be produced by most melt-flow plastics processes, but it is a difficult shape to produce by rotational molding. A hollow rib is an ideal stiffening rib shape for rotational molding (Fig. 3.3B). In the final analysis, the best part design is the one that adapts to the process without exceeding its capabilities. The part design details that are suitable for rotational molding are presented in this chapter.

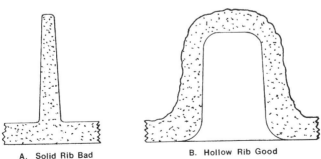

A. Solid Rib Bad   B. Hollow Rib Good

Figure 3.3 The solid reinforcing rib (A) is not an acceptable shape for rotational molding, which is at its best producing hollow stiffening ribs (B) that provide greater strength

3.3   Design for the Material

Rotational molding is a material-dependent plastics part manufacturing process. The high oven temperature and lack of melt-flow pressures limits the type of plastics that are suitable for the process. These materials, in turn, impose limitations on the shape of the parts that can be produced. For example, nylon requires larger corner radii than PE. PVC can be molded with smaller draft angles than PP. It would not be unusual to discover that a product being produced in LLDPE could capture a new market if its operating temperature could be increased. The logical answer to this opportunity would be to change to HDPE, PP, or nylon. These higher temperature, stiffer materials might not be moldable in the same undercut mold that produced the LLDPE parts. The design guidelines will vary for different plastic materials. Where these differences are important, they are defined in the design guidelines presented in this chapter.

3.4   Part Design

During the part design phase of the project, the designer concentrates on the multiplicity of interrelated details that combine to make a whole part. The product design requirements for a chair armrest, a floor-scrubber tank, and a tractor instrument panel are all very different. The individual part design details for these three products will, however, be exactly the same. The plastic material and the process do not recognize the differences between an instrument panel and an armrest. They only know that the requested shape can or cannot be produced. Designing for the process and material can be the difference between success and failure. Fortunately, these design details for rotational molding have been evolved and are now highly refined.

3.4.1   Wall Thickness

The thickness of a rotationally molded part is dictated by two primary considerations. The wall thickness must provide for the functional requirements of the product, while accommodating the molding requirements of the process. A flexible medical waste collection bag might be strong enough to function properly with a 0.5 mm (0.020 in.) wall thickness, but the process can only

Table 3.1   Recommended Wall Thickness for Commonly Rotationally Molded Materials

Plastic material	Ideal		Possible	
	Min. mm (in.)	Max. mm (in.)	Min. mm (in.)	Max. mm (in.)
Polyethylene	1.50 (0.06)	12.70 (0.50)	0.50 (0.02)	50.80 (2.00)
Polypropylene	1.50 (0.06)	6.40 (0.25)	0.75 (0.03)	10.16 (0.40)
Polyvinyl chloride	1.50 (0.06)	10.16 (0.40)	0.25 (0.01)	25.40 (1.00)
Nylon	2.50 (0.10)	20.32 (0.80)	1.50 (0.06)	31.75 (1.25)
Polycarbonate	2.00 (0.08)	10.16 (0.40)	1.50 (0.06)	12.70 (0.50)

produce that shape with a 1.0 mm (0.040 in.) thickness. In this case, processing would take priority over function.

The ideal wall thickness is the thinnest wall that will provide for both the functional and the processing requirements of the product. Thickness has a direct effect on cost. Plastic material represents a significant fixed cost that cannot be influenced by the molder or the customer. Thinner walls reduce both material cost and molding cycle time. Generally speaking, thinner is better.

The minimum allowable thickness is determined by strength requirements and the material's ability to uniformly coat the cavity. The maximum allowable wall thickness is dictated by cycle time and the material's ability to withstand high temperatures without degradation. The wall thicknesses that are suitable for the commonly molded materials are shown in Table 3.1.

One of the advantages of rotational molding is that once the mold has been built, it can be used to produce parts with thicker and thinner walls without mold changes, by simply charging the mold with more or less material. The optimum wall thickness can then be established by testing the actual part. This test is always more reliable than strength calculation or speculation. There are few other processes that provide the designer with this capability.

Rotational molding is not the ideal process for producing a part that requires variation in wall thickness. The only material that lends itself to wide variations in wall thickness is PVC. A special molding technique called *stop rotation* allows some parts, such as the anesthesia face mask (Fig. 2.8), to be produced with both thick and thin walls. This molding technique stops the rotation of the mold in a specific position after the PVC has coated the cavity. Gravity then causes the ungelled material to drain into the lowest part of the cavity.

Rotational molding is an open molding process that defines only those surfaces of a part that are in contact with the cavity. The inside surfaces are free-

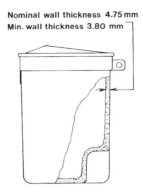

Nominal wall thickness 4.75 mm
Min. wall thickness 3.80 mm

Figure 3.4 Wall thickness should be specified as average and minimum allowable. See Table 3.1

formed. Once the mold is built, it does not change. The wall thickness is thereafter controlled by the amount of material put into the mold and the cycle-to-cycle variations of the process.

The exact wall thickness of a rotationally molded part cannot be specified in the common manner employed for closed-molding processes, such as injection and compression molding. The ideal way to specify a wall thickness for this process (Fig. 3.4) is to indicate both the nominal, or average, wall thickness and the minimum thickness that is acceptable anywhere on the part.

Wall thickness uniformity is dependent on the size and shape of the part and the material being molded. A commercial thickness variation is ±20%. Thickness variations of ±10% can be achieved in some cases where uniformity is more important than cost. The wall thickness variations do not include the thickening of walls at outside corners, as discussed in Section 3.4.3.

3.4.1.1   Warpage

During the cooling portion of the molding process, the plastic material contracts or shrinks. Shrinkage of the hollow parts allows them to pull away from the cavity before the material has cooled enough to be strong enough to retain its shape. This allows large, flat surfaces to warp (Fig. 3.5). The industry-established standards for warpage are shown in Table 3.2.

Figure 3.5 For flat-panel warpage specifications see Table 3.2

Table 3.2   Flat-Panel Warpage Standards for Commonly Molded Materials in±cm/cm and in./in.

Plastic material	Ideal	Commercial	Precision
PE	0.050	0.020	0.010
PP	0.050	0.020	0.010
PVC	0.050	0.020	0.010
Nylon	0.010	0.005	0.003
PC	0.010	0.005	0.003

Ideal = No extra care required
Commercial = Requires special care
Precision = Available at added cost

This type of shrinkage-related warpage can be significantly reduced by pressurizing a hollow part during the cooling part of the molding cycle. This pressure forces the part to cool while being held in contact with the cavity. Forcing the part into contact with the cool cavity has the added benefit of reducing the time required to cool the part. The use of pressurized air or inert gas has many benefits, but not all molding machines are equipped for this type of molding.

A simpler approach to discouraging the warpage of large, flat surfaces is to avoid designs of that type. If flat surfaces cannot be eliminated, they can be strengthened to resist warpage with the use of ribs, steps, crowns, and domes (Fig. 3.6). In this case, a large, round, flat-topped tank has been redesigned to resist warpage. A dome as small as 0.015 cm/cm (0.015 in./in.) is enough to discourage warpage, but the larger the doming or crowning, the less warpage there will be.

Figure 3.6 Domed, crowned, stepped, and ribbed surfaces discourage and disguise warpage

Highly polished surfaces that reflect light exaggerate the appearance of a warped surface. Textured surfaces tend to disguise warpage.

3.4.1.2   Parallel Walls

Rotational molding excels at producing hollow parts with closely spaced parallel walls, such as the cross-section of the boat shown in Fig. 1.5, where the inner and outer hulls are molded as one large, integral part. Insulated food containers,

Figure 3.7 One-piece, rotationally molded, double-walled flower pot with closely spaced parallel walls

ice chests, tote bins, and cushioned shipping cases all rely on this technique. The small double-walled flower pot shown in Fig. 3.7 uses this capability to produce a one-piece structure. Most other plastic processes would produce this flower pot as two parts requiring assembly (Fig. 3.8).

Closely spaced parallel-walled parts, such as the boat and the flower pot, present two problems. When the depth of the recess is greater than the width across the open end of the part, it becomes difficult to adequately heat the mold at the inside bottom corners. If these locations on the mold are slow in reaching molding temperature, the molded parts will contain thin walls in those areas. The shallow depth-to-width ratio of the boat minimizes this potential molding problem. There are methods for forcing heat into these inside bottom corners, but they are all costly. The best approach is to avoid designing parts with these proportions. When this situation cannot be avoided, this undesirable detail should be reviewed with a knowledgeable molder before finalizing the part design.

The second problem associated with closely spaced parallel walls is providing enough volume in the cavity for the plastic material. Finely ground powdered plastic has a bulk density factor at least three times greater than the same material molded into a solid part. The distance between two parallel walls must provide enough volume for the powdered plastic, and enough space for the material to flow through the cavity and uniformly coat the cavity walls. Flat panels with closely spaced parallel walls, of the type required for a cart door, table top, or business machine housing panel, are produced in cavities that

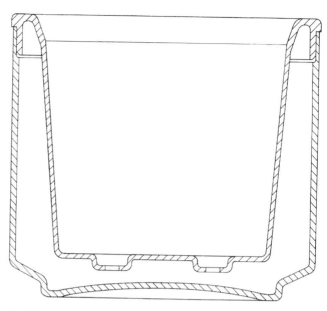

Figure 3.8 Double-walled flower pot assembly with two injection-molded parts

severely limit the space available for the powder. The cross-section of the part shown in Fig. 3.9 has parallel walls that are too close together. In this case, the plastic has partially bridged the gap between the two walls. The resulting part is no longer completely hollow. Once these bridges are formed, they restrict the free flow of the plastic powder through the cavity. The plastic material bridging the gap was intended to be somewhere else on the part. This condition requires the use of more plastic material to avoid a thin wall at some other critical location on the part.

The refuse container (Fig. 3.10) has more than enough volume in the cavity to accept the full charge of plastic powder. There are, however, powder flow problems in the closely spaced walls that form the pickup truck lifting hook recess and the handles.

The absolute minimum distance between two walls must be three times the nominal wall thickness for the efficient molding of good quality parts. A distance of five times the nominal wall thickness is desirable. The lifting hook recess, shown in Section X-X, is relatively open and easy for the powdered plastic to reach and coat.

The handle, as shown in Section Y-Y, is much more restrictive to powder flow. In structures of this type, the minimum distance between walls must be five times the nominal wall thickness.

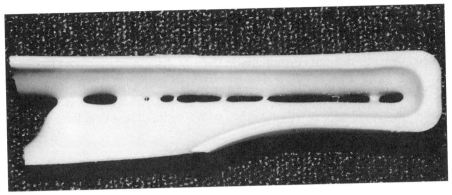

Figure 3.9 Closely spaced parallel wall with undesirable bridging

This handle could have been designed to extend all the way across the container, but the center support adds strength to the handle. Of equal importance is that this hollow handle support, as shown in Fig. 3.1, provides a third opening for powder to flow into the handle while reducing the length it has to travel.

The rotational molding process is noted for its ability to produce hollow plastic parts with uniform wall thicknesses. The process is not good at producing parts that require abrupt changes in wall thickness. Some gradually changing wall thickness can be produced by changing the thermal conductivity of the mold in specific areas. Welding an aluminum panel into a fabricated steel mold would increase the thermal conductivity of the cavity in that location. The aluminum surface of the cavity would reach molding temperature before the steel surfaces. The aluminum surfaces would then have a longer time to pick up the plastic powder than the steel surface would. The same effect can be achieved by varying the wall thickness of the cavity. In other instances, heat-absorbing projections are placed on the outside surface of a cavity in an area where a thicker wall is desirable. The reverse effect can be achieved by shielding or insulating portions of the cavity so that they take longer to come up to molding temperature. These techniques and others extend the capabilities of the process, but rotational molding is at its best producing hollow parts with uniform wall thicknesses.

A properly designed part and a good-quality mold that heats uniformly will produce parts with a uniform wall thickness. This is highly desirable, as parts containing thick walls take longer to form and to cool. The plastic in thick walls stays hot longer and shrinks more than that in thin walls, which cools faster. A molded part with both thick and thin walls will have different shrinkage factors in different locations. These differences in shrinkage create molded-in residual stress and a propensity for postmold warpage.

The nonuniform cooling associated with variations in wall thickness also affects the percentage of crystallinity in the molded part, as reviewed in Section 2.6.2.

Referring to the refuse container in Fig. 3.10, if the lifting hook recess and the handle cross-sections were specified with a width of less than three times the

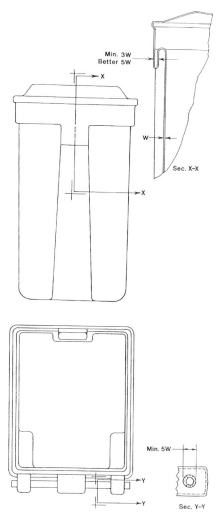

Figure 3.10 Minimum spacing guidelines for closely spaced parallel walls

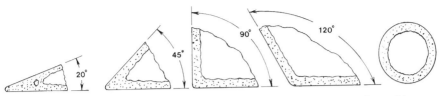

Figure 3.11 Corner angles of less than 45° produce thicker walls, voids, and increased shrinkage

nominal wall thickness, the molding process would try to produce a solid part in these locations. The powder would not be able to penetrate into these restricted areas and the process would produce a partially filled solid section. The thicker sections would stay hot longer and shrink more than the rest of the part. This would encourage warpage and molded-in stress. Molded-in stress will weaken the lifting hook recess, which needs as much strength as possible. These two details could be molded as solid sections by processes such as injection-molded structural foam or reaction injection molding, but they should be designed as hollow sections for the rotational molding process.

Another consideration that affects the flow of the powder and wall thickness uniformity is the angle between two intersecting walls [17]. As the angle between two walls becomes less than 90°, the open space between them is reduced (Fig. 3.11). With angles of 45° or less, the two walls begin to act like closely spaced parallel walls. These converging walls violate the minimum allowable space between parallel walls before they meet at the corner of the part. This makes it difficult, or impossible, for the powdered plastic to flow all the way into the corner. These corners are difficult to coat uniformly, and they often contain internal voids, as shown in the part in Fig. 3.12.

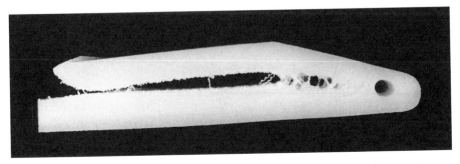

Figure 3.12 Small corner angle causing bridging, a thick section, and internal void

Nylon has been successfully molded into angles as small as 20°. PE and PVC can have problems with angles of 30° or less. PC is better with angles of 45° or greater. All the commonly moldable materials can accommodate a 90° or greater angle. The ideal shape for rotational molding would be a sphere, which has no corners at all.

In situations where a small angle is required, these problems can be minimized by keeping the extension short and providing the largest allowable radius at the corner where the two walls meet.

Angles of less than 45° accumulate more material than larger angles. The resulting thicker sections take longer to cool. The additional shrinkage in these thick corners contributes to molded-in stress and warpage.

3.4.2 Reinforcing Features

The cost of a rotationally molded part is, to a great extent, dictated by the plastic material being molded and the wall thickness of the part. The ideal wall thickness is always the thinnest wall that will satisfy both the functional and manufacturing requirements of the product. Rotational molding excels in the production of large parts with relatively thin walls. During the rotational molding process, the plastic material simply adheres to and coats the cavity as the molding machine rotates the hot cavity through the puddle of material in the bottom of the cavity. The molten plastic material does not have to flow through the length of the cavity. In many instances, the process is capable of molding parts with walls too thin to satisfy the functional requirements of the product. In these cases, the wall thickness must be increased, and function takes priority over cost and processing considerations.

Increasing the wall thickness will produce a stronger part. There are other ways of increasing strength, while keeping the wall thickness to a minimum. For a given wall thickness, radiused corners are stronger than square corners. Crowning or doming a flat surface increases its stiffness (Fig. 3.6). The most frequent technique for increasing the strength of a thin-walled part is the use of reinforcing ribs. The multiple reinforcing ribs on the underground septic tank (Fig. 3.13) provide this relatively thin-walled part with strength enough to resist the crushing pressure of the soil that surrounds it.

Rotational molding is not a good process for producing the common solid reinforcing ribs used on parts produced by closed-molding techniques. This process is at its best while producing hollow ribs (Fig. 3.3).

Figure 3.13 Stiffening ribs used to provide strength on heavily loaded underground septic tanks (Courtesy Association of Rotational Molders)

The refuse container (Fig. 3.14) is rectangular in cross-section. The four flat walls on the sides of the container are subject to an outward thrusting force when the container is filled to capacity. Adding vertical reinforcing ribs on these four walls would increase their ability to resist that force. Doming these surfaces inward or outward would also increase their stiffness. The bottom of the container, and if necessary the lid, could also be stiffened with reinforcing ribs.

The shape of reinforcing ribs can be rounded or trapezoidal in cross-section. The ribs must project above or below the nominal wall of the part a distance of at least four times the nominal wall thickness, in order to provide a significant stiffening effect.

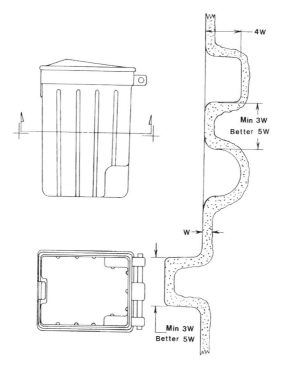

Figure 3.14 Minimum proportions for stiffening ribs and inter-rib spacing

Hollow reinforcing ribs become closely spaced parallel walls, and they must follow the same rules. Outwardly projecting hollow ribs must have a minimum width of three, and preferably five, times the nominal wall thickness.

If the width of a rib, or the space between two ribs, is less than three times the nominal wall thickness, the powdered plastic has difficulty flowing into these restricted areas. This problem becomes magnified as the depth of these recessed areas increases. Ribs that project more than four times the nominal wall thickness should be designed with proportionately wider spaces for the powdered plastic to flow into.

Solid reinforcing ribs are not recommended for rotationally molded products. In those cases when solid ribs cannot be avoided, they must be kept as small as possible. The proportion of solid ribs that have been successfully produced are shown in Fig. 3.15 [18]. Ribs of this type can only be produced as outwardly extending projections.

These solid ribs are thicker and take longer to cool than the nominal wall of the part. Solid ribs will shrink more than the rest of the part. The increase in shrinkage in these thicker and stronger solid ribs may deform the walls to which

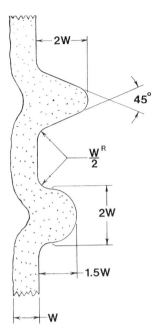

Figure 3.15 Shallow, solid stiffening ribs are sometimes possible, but are not recommended

they are attached. The increase in shrinkage in solid ribs normally results in a sink mark on the inside surface of a part. These sink marks may or may not be acceptable.

It is a common practice to provide additional strength by attaching two closely spaced parallel walls to each other. This technique can convert two relatively weak walls into one integral box-beam structure that is inherently strong. The thickness of the wall in the kiss-off area (Fig. 3.16) is almost always established by trial and error, but 1.75 times the nominal wall thickness should be specified on the drawing as a starting thickness. This spacing provides room enough for the powdered plastic to move freely through the cavity. The kissing off of the two walls takes place only as the last of the powdered plastic coats the cavity.

Kiss-off reinforcing ribbing refers to long, continuous kiss-off areas. *Tack-off* reinforcing refers to interrupted kiss-off areas. These tack-offs can be any shape, but round is the most common (Fig. 3.17).

In those instances where kiss-off ribbing or tack-offs are undesirable on an appearance surface, the kiss-off can be designed into only one wall of the part on the nonappearance side.

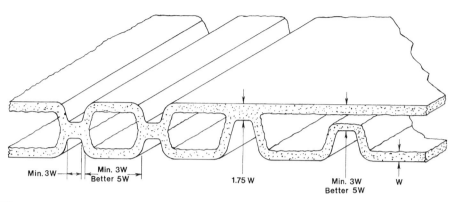

Figure 3.16 Proportions for various types of kiss-off and almost kiss-off reinforcing ribs

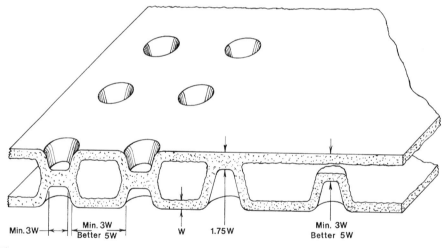

Figure 3.17 Proportions for various types of tack-off and almost tack-off reinforcements

Tack-off and kiss-off reinforcing are sometimes used in rectangular portable liquid tanks to strengthen the side walls and discourage the liquid in the tank from sloshing from side to side (Fig. 3.18). In these situations, there can be a substantial distance between the two walls that are to be attached to each other. In these cases, kiss-off ribbing should be designed into both walls. This minimizes the depth of the recess and makes it easier to heat the mold in the actual kiss-off area at the center of the tank.

Kiss-off ribbing has been used in the bottom of double-walled tanks for products such as insulated ice-making machines. In these applications, the kiss-

Figure 3.18 Center kiss-off rib provides strength and discourages side-to-side sloshing of fuel in this nylon tank

off strengthens the inner wall. Failures can develop at the edge of the kiss-off on tanks that hold liquids or products such as grain that act like liquids. These failures have been traced to the added strength at the kiss-off. The inner bottom wall of the tank between the kiss-offs bends under the load. Stresses build up at the junction between the weaker inner bottom wall and the stronger kiss-off. In some cases, a more durable tank has been produced with what is referred to as an *almost kiss-off*. An almost kiss-off brings the inner and outer walls of the tank close together, but they are not attached to each other. As the inner wall is loaded, it bends and comes to rest on the almost kiss-off. This supports the inner wall, while leaving it free to move relative to the kiss-off. Kiss-off and almost kiss-off details are shown in Figs. 3.16 and 3.17.

The inner and outer hulls of the boat in Fig. 1.5 have kiss-off ribbing along the keel. Almost kiss-off ribs have been provided along the outside edges of the deck. If the deck flexes under load, it will gain strength from the outer hull, which is supported by the water. At the same time, the inner and outer hulls are free to move relative to each other in response to inside or outside forces.

There are large quantities of open-topped tanks, drums, and containers produced by rotational molding. It is a common practice to mold a long,

Figure 3.19 Various approaches to stiffening the open end of a thin-walled, cylindrical tank

cylindrical, or square part that is cut in the center to produce two shorter containers open at one end (Fig. 6.6). It is even more common to mold a tank of the required size and then cut an opening in one end (Fig. 3.19A).

Removing one end of the tank produces a weak side wall at the open end (Fig. 3.19B). That wall could be strengthened by increasing the thickness of the whole tank (Fig. 3.19C), but that would be a waste of material and cycle time.

In some instances, it is possible to strengthen the top edge of the tank by foaming the same amount of plastic material to produce a thicker and stiffer wall.

The top of the tank could also be stiffened by incorporating an inward or outward projecting rib just below the top of the tank (Fig. 3.19D).

A thin-walled tank with the same strength could be produced by removing the top wall, so as to leave an inward-projecting flange on the part (Fig. 3.19E). This flange could provide even more strength if it also extended down into the tank (Fig. 3.19F).

If an inward projecting flange is undesirable, the flange can extend outward (Fig. 3.19G). This structure would be even stronger if that flange also extended downward (Fig. 3.19H).

An extremely strong tank top rim could be provided by creating a hollow box-beam flange, using the kiss-off rib technique (Fig. 3.19I).

All the configurations shown in Figs. 3.19D through 3.19I could be used to produce a low-cost, thin-walled part, with added strength at the open end. Some of these stiffened flanges are more complex than others, but all of these shapes could be produced with a straight opening and closing two-piece mold.

3.4.3   Corner Radii

The rotational molding process is at its best producing hollow parts with smoothly blended contours. Providing radii on the corners of these parts has three major benefits. The corners of a product are frequently heavily loaded. Radii distribute these loads over a broader area, resulting in a stronger part. This is especially true for parts that have to withstand high impact loads. It is sometimes possible to produce a stronger, lower cost part by increasing the size of the radius while reducing the nominal wall thickness. Larger corner radii improve the flow of the plastic powder through the sometimes complex contours of the cavity. Powdered and liquid plastic materials can easily flow across the surface of a corner with a large radius. It is more difficult for the material to flow into a sharp corner. As a powdered material flows through a sharp corner, it hesitates and stops moving until the mold has rotated enough for the material to fall away from the corner. This hesitation of the material contributes to the increase in wall thickness that is common with square outside corners.

Large corner radii heat up to molding temperatures more uniformly than corners with small radii. The cross-section of the refuse container body shown in Fig. 3.20 illustrates this phenomenon. The small outside corners to the right accumulate more material than the corners with the larger radii on the left. These small outside corners on the mold are actually being heated from two sides and they quickly come up to molding temperature. The corners of the mold with the large radii come up to molding temperature at closer to the same time as the surrounding flat surfaces of the mold.

The outside corners on a mold are closer to the source of the oven heat than the inside corners. By the time the inside corners reach molding temperature, the outside corners have already accumulated more than the intended amount of plastic material. This can result in sharp inside corners with walls that are

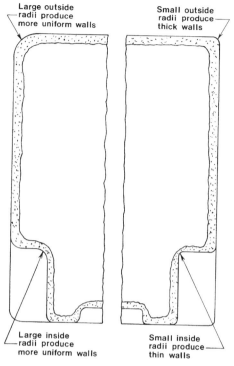

Large outside
radii produce
more uniform walls

Small outside
radii produce
thick walls

Large inside
radii produce
more uniform walls

Small inside
radii produce
thin walls

Figure 3.20 Large corner radii improve uniform heating of the mold resulting in less variation in wall thickness

Figure 3.21 Outside corners are typically thicker and stronger than inside corners

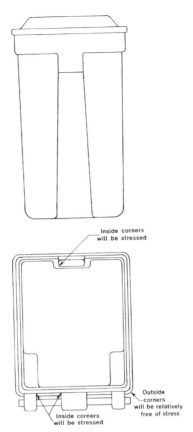

Figure 3.22 Confined inside corners that shrink onto cores will contain more stress than outside corners that are free to shrink away from the mold. See Table 3.3

thinner and weaker than the nominal wall. This condition can be seen on the part in Fig. 3.21. Providing large inside corner radii minimizes this variation in wall thickness.

The increased mass of material in sharp outside corners stays hot longer and shrinks more than the thinner inside corners and the nominal wall. This nonuniform shrinkage creates molded-in stress and encourages warpage.

Many molded parts contain both inside and outside corners. Rotationally molded parts are produced in open molds with no internal cores. Outside corners are free to pull away from the cavity as the part cools and shrinks. The plastic material in these corners is free to shrink, and these corners will be virtually free of stress.

Table 3.3   Recommended Radius Size for Commonly Molded Materials

Plastic material	Outside radii		Inside radii	
	Min. mm (in.)	Better mm (in.)	Min. mm (in.)	Better mm (in.)
PE	1.52 (0.060)	6.35 (0.250)	3.20 (0.125)	12.70 (0.500)
PP	6.35 (0.250)	12.70 (0.500)	6.35 (0.250)	19.05 (0.750)
PVC	2.03 (0.080)	6.35 (0.250)	3.20 (0.125)	9.53 (0.375)
Nylon	4.75 (0.187)	12.70 (0.500)	6.35 (0.250)	19.05 (0.750)
PC	6.35 (0.250)	19.05 (0.750)	3.20 (0.125)	12.70 (0.500)

Inside corners, such as those on the handle supports and around the lifting hook recess (Fig. 3.22), are formed over a core pin. The core pin prevents the plastic material around the core from shrinking the normal amount. All the inside corners in these areas will contain molded-in residual stress. The larger these inside corner radii are, the lower the stress will be. This situation explains why inside corner radii on rotationally molded parts should be larger than those required on outside corners.

The recommended inside and outside corner radii for the commonly molded materials are listed in Table 3.3. As far as radii are concerned, bigger is better. The ideal corner radius for a rotationally molded plastic part is the largest radius that the functional requirements of the product will allow.

3.4.4   Molding Draft Angles

Molding draft angles are tapers that are provided on those surfaces of a part that are perpendicular to the parting line of the mold. The function of draft angles is to improve the release of the part from the mold. The cooling portion of the molding cycle is controlled in part by the time required to cool a molded part to the point that it has regained strength enough to retain its shape while being forced out of the mold. Draft angles reduce the force applied to a part during demolding. The liberal use of draft angles can result in a reduced cooling cycle and a lower part cost.

Rotational molding is an open-molding process. As a hollow plastic part cools and shrinks, it has a tendency to pull away from the cavity. This allows some parts in some materials to be molded without draft angles. Design engineers sometimes choose rotational molding over other processes because of

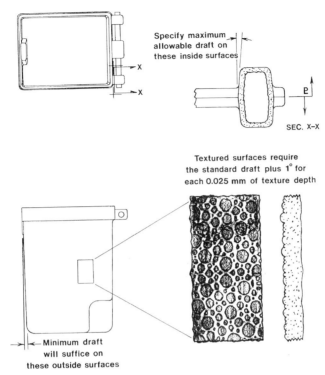

Figure 3.23 Molding-draft angle considerations on inside, outside, and textured surfaces. See Table 3.4

its ability to produce parts with straight walls. Eliminating draft angles can be a distinct advantage in some applications.

The body portion of the refuse container (Fig. 3.23) is free to pull away from the cavity as it cools and shrinks. The draft angle on these outside surfaces can be minimized.

The plastic material in the wall between the handle supports and the handle supports themselves is not free to shrink as much as the rest of the body of the container. These inside surfaces are restricted from shrinking by the metal cores of the mold between the two handle supports. These surfaces require larger draft angles.

Many rotationally molded parts can be quite large in size. Even a small draft angle has a noticeable effect on the size and shape of large parts. Draft angles should be specified by the designer and incorporated into the part drawing, in order to determine their effect on a part's geometry and overall appearance. It is never a good idea to leave this detail up to the mold maker.

Specifying the optimum draft on a rotationally molded part requires careful consideration. Each plastic material has its own unique requirements. The soft PEs and PVCs can be easily demolded from cavities containing shallow undercuts, which are negative draft angles. The rigid PPs, nylons, and PCs cannot accommodate undercuts.

The high mold shrinkage factor materials that are free to shrink away from the cavity, will shrink away from minor undercuts, such as those formed by tool marks, welded seams, and textured and shot-blasted surfaces. PE's high shrinkage is useful in these instances. PC, which has a low mold-shrinkage factor, does not pull away from the cavity enough to accommodate these small undercuts.

It is only logical that rigid plastic materials with low mold-shrinkage factors, will require larger draft angles than the softer materials with higher shrinkage factors.

Parts molded in rigid plastic materials will be easier to demold from cavities with large draft angles and smoothly polished surfaces. Cavity polishing can increase the cost of a mold. This added cost can be minimized by specifying larger draft angles.

The recommended inside and outside surface draft angles for the commonly molded plastic materials are shown in Table 3.4. These draft angles have been found to be acceptable in the majority of cases, but bigger is better, and there are always exceptions.

The size and shape of a part and the material being molded can combine to require special draft angle considerations. The portable storage chest (Fig. 3.24) was molded as a double-sided part with closely spaced parallel walls. The inner walls incorporate five drawer-supporting recesses on three sides of the part. The inside surfaces that form the recesses will shrink onto the cores that form them. Shrinkage of the plastic onto the core over such a large area will render this part

Table 3.4   Recommended Draft Angles for Commonly Molded Materials in Degrees per Side

Plastic material	Inside surfaces		Outside surfaces	
	Min.	Better	Min.	Better
PE	1.0°	2.0°	0.0°	1.0°
PP	1.5°	3.0°	1.0°	1.5°
PVC	1.0°	3.0°	0.0°	1.5°
Nylon	1.5°	3.0°	1.0°	1.5°
PC	2.0°	4.0°	1.5°	2.0°

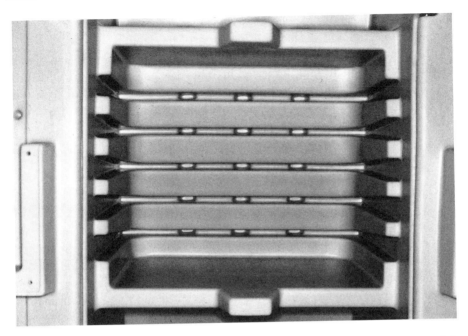

Figure 3.24 The drawer supports on this tool storage chest shrink onto the core of the mold making demolding difficult (Courtesy Trilogy Plastics, Inc., Louisville, OH)

difficult to remove from the mold. The larger than normal draft angles shown on these recesses reduce the force required to remove this part from the mold. Details such as these recesses, which complicate the molding of a part, should be reviewed with a knowledgeable molder before finalizing the design of a part.

3.4.5   Texturing Draft Angles

Special finishes are specified on the surface of rotationally molded parts in order to change their natural, as-molded appearance. Polished, textured, shot-peened, and abrasive grit blasted finishes are all used to hide surface imperfections and to make a part look like some other material, such as leather or wood. Shot-peening is one of the most frequently used finishes on rotationally molded parts. Textured, blasted, and peened surfaces (Fig. 3.23) are formed by a multiplicity of small undercuts. These small undercuts require larger draft angles in order to easily demold a part. Stiff, high-shrinkage materials cause the most difficulties on textured inside surfaces. The reverse is true on outside surfaces. As a general rule, it is highly desirable to avoid these special finishes on inside surfaces where the plastic material will shrink onto the cores of the mold.

The minimum allowable draft angle on outside textured, peened, and blasted surfaces is the standard draft indicated in Table 3.4, plus one additional degree per side. The ideal draft angle in these areas would be the standard draft plus one degree and one additional degree for each 0.025 mm (0.001 in.) of finish depth.

3.4.6   Molded Undercuts

The use of undercuts on rotationally molded parts is always a compromise between providing a desirable functional feature and keeping the tool-building and molding operations simple. Anything that causes tool-building and molding difficulties increases costs. These added costs are sometimes acceptable, because undercuts can provide desirable features such as stiffening ribs, snap-fit surfaces, threads, pressure pads, and standoffs.

An undercut is any inward- or outward-projecting surface that is parallel to the mold's parting line. The imaginary round part shown in cross-section in Fig. 3.25 has seven different kinds of undercuts. It is difficult to imagine a use for such a part, but if a part of this shape were required, its seven undercut surfaces would require the following considerations.

The shape of the part to be produced and the location of the mold's parting line determine which surfaces are and are not undercuts. As a round part, the lowest cost mold would be one with a parting-line location at PL 1. With the parting line in this location, all seven of the inward- and outward-projecting surfaces would be undercuts.

If the parting line of the mold was placed at PL 2, undercuts 1 through 6 would cease to be undercuts. The mold would cost more, but the molding of the parts would be simplified. The PL 2 parting-line location would also produce a parting-line scar the full length of the part. If there was any flash at the parting line, it would have to be trimmed all the way along a complex surface. Flash would also interfere with the function of the threads (1).

It is interesting to note that if the part was molded in a flexible PVC, it could be collapsed with a vacuum and demolded through the small opening at PL 3. An electroformed cavity of this type would be capable of producing this part with no parting-line or flash-trimming scars on the primary appearance surfaces of the part.

With the parting-line location at PL 1, wall A must deform inward in order to release undercuts 2 through 5 from their recesses in the cavity.

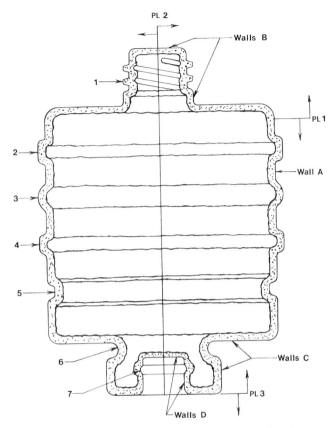

Figure 3.25 Various types of rotationally molded undercuts

As the main body of the part is pulled out of the cavity, the shape of the undercut must be such that it will force wall A inward. The shape of undercut 2 would not encourage A to deform inward. The rounded shape of undercut 3 would develop an inward-thrusting force. The 45° angles on undercuts 4 and 5 are of the ideal shape to encourage wall A to deform inward.

In order to free undercuts 1 that form the threads, it would be necessary to deform walls B. These circular walls are smaller than the long wall A. These walls will be difficult to deform enough to free the thread undercuts. An alternative approach would be to unscrew the threads from the cavity with a threaded insert in the mold.

Undercut 6 suffers from a similar problem in that the shape of walls C makes them structurally strong. A high force would be required to deform walls C.

The ability of walls A, B, and C to deform enough to free these undercuts is dependent on the thickness of the walls and the stiffness of the plastic material.

A 3.8 mm (0.150 in.) wall will deform more easily than a 10.2 mm (0.400 in.) wall. An LLDPE part will deform with less force than that required for a part molded in an HDPE material.

It should also be pointed out that if this part was demolded through PL 1, the inward-projecting undercut 5 would be in the best location. If this undercut was located just below PL 1, it would require additional deformation of walls A. The inward-projecting metal that forms undercut 5 reduces the diameter that undercuts 2, 3, and 4 would have to pass through.

If the distances between undercuts 2, 3, and 4 were the same, undercuts 3 and 4 would expand into undercut recess 2 as the part was pulled out of the cavity. Wall A would have to deform more than once for the part to be demolded. This potential problem can be avoided by varying the distances between undercuts 2, 3, and 4.

In the final analysis, the best location for the parting line for this mold would be at PL 2.

Undercuts 1 through 6 are outside undercuts on a hollow part. The normal shrinkage of the material will help release these undercuts from their recesses in the cavities. If this round part were 100 cm (39.4 in.) in diameter at wall A, the shrinkage of an HDPE material would reduce that diameter by 2.24 cm (0.885 in.). Theoretically, shrinking would free an undercut with a depth of 1.12 cm (0.44 in.) per side. Regrettably, it takes many hours for an HDPE to shrink the full amount. With a part of this size, it would be practical to assume that at least half the shrinkage would have taken place by the time the part was ready to be demolded. Shrinkage could be relied upon to free undercuts whose depth was approximately 0.56 cm (0.22 in.) per side. Undercut 6 is attached to a wall with a diameter of only 50 cm (19.7 in.). The shrinkage of this smaller diameter can be relied upon to free an undercut that is only half as deep as those attached to wall A on a diameter than is twice as large.

The shrinkage of a plastic material during the cooling portion of the molding cycle is helpful in freeing undercuts 1 through 6 from the mold. This same shrinkage is detrimental to undercut 7. This undercut is formed over a core pin that restricts the material from shrinking. Such shrinkage as there is causes the material to grip the core pin, which will make undercut 7 difficult to remove from the mold. Walls D must stretch in order to allow the undercut to be freed from the mold. Deep inside undercuts of this type cannot be produced in rigid materials such as HDPE, PP, nylon, and PC.

The undercut formed by the lid on the refuse container (Fig. 3.26) would be freed from the cavity by shrinkage of the plastic material. Using shrinkage in this beneficial manner allows this relatively complex part to be produced in a simple two-piece mold.

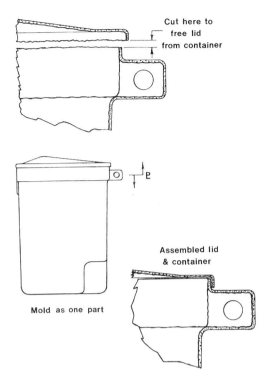

Figure 3.26 Mold shrinkage will free the small undercut formed by the outward-projecting lid on this refuse container

The presence of undercuts always complicates molding and tool construction. The benefits provided by undercuts often outweigh these difficulties. The size, shape, location, and material being used to mold undercuts must always be carefully thought out and executed.

3.4.7   Molded Holes

During the heating portion of the rotational molding process, the plastic material coats all hot surfaces on the cavity that it comes in contact with. One advantage of this process is that the molded parts do not contain the weld-lines that weaken parts produced by the melt-flow processes.

Rotational molding is not ideal for producing parts with holes through the wall. This is a handicap that the process shares with thermoforming and blow

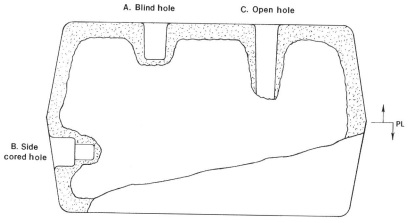

Figure 3.27 Typical inward-projecting round holes. A and C are inline holes. B requires a side-acting core pin

molding. In spite of this limitation, molders and tool makers have succeeded in developing techniques for molding holes through, into, and onto rotationally molded parts. Every conceivable size and shape of hole and recess has been molded, but round holes are the most common.

Holes that project into a molded part are the easiest to produce (Fig. 3.27). They are formed when the plastic coats inward projecting core pins. The diameter and depth of these holes are only limited by the process's ability to heat the core to a high enough temperature for the plastic to adhere to and coat the core pin. A length-to-diameter ratio of four to one is possible with small, solid steel cores. A larger ratio can be achieved by using core pins made of higher heat conducting aluminum and copper alloys.

Blind holes (Fig. 3.27A and B) are produced when the plastic coats the free end of the core pin. An open, or *through* hole (Fig. 3.27C) can be molded when the core pin extends into the cavity so far that the free end cannot be heated enough to be coated by the plastic material. The free ends of these cores are sometimes coated with slippery baked-on fluorocarbon or silicone that makes it difficult for the plastic to adhere. Some open holes are produced by machining a hole in the bottom of a blind hole.

Holes (Fig. 3.27A and C) are positioned perpendicular to the mold's parting line. The core pins that form these holes are withdrawn from the molded part in the normal mold-opening operation. The hole (Fig. 3.27B) on the side wall of the part is formed by a core pin that is parallel to the parting line. This core pin must be withdrawn from the cavity before the part can be demolded. Side-cored holes of this type are widely used, but they do require more effort on the part of the machine operator.

Outward-projecting holes are more difficult to produce. The outward-projecting open and closed holes (Fig. 3.28A and B) are single-walled structures that cannot be produced by rotational molding. The liquid PVC plastisols are the only plastic material that can be considered for single-wall details of this type.

An outward-projecting open hole (Fig. 3.28C) can be produced by molding a closed, hollow, tubular projection that is then cut to length after demolding. Hollow projections of this type can be machined to provide inside or outside threads. A flexible hose can be clamped or welded to the projection. Round projections of this type require a minimum outside diameter of at least five times the nominal wall thickness in order to mold properly.

An outward-projecting closed hole, or blind boss, such as that shown in Fig. 3.28B, could easily be provided by closed-molding processes such as injection, compression, or structural foam molding. Blind bosses are frequently used with threaded fasteners to locate and anchor a tank, or for the mounting of pumps or

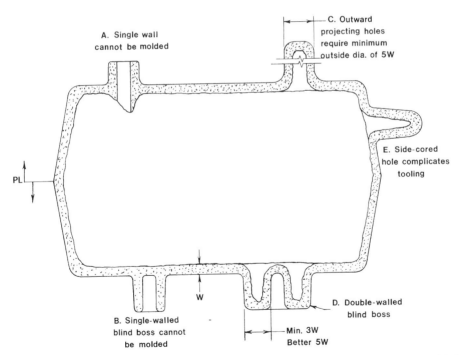

Figure 3.28 Outward-projecting round holes. Solid walls at A and B cannot be molded. Hollow structures around holes C, D, and E are moldable. Hole E requires a side-acting cavity section

motors. If a blind boss is required on a rotationally molded part, it must be designed with enough space around the core pin to accommodate the flow of the powdered plastic material. The walls around such a hole are closely spaced parallel walls. The open space for the plastic material must be a minimum of three and preferably five times the part's nominal wall thickness, as shown in Fig. 3.28D.

The holes in Fig. 3.28A through D are in line with the opening of the mold. These holes can be provided with a simple two-piece mold. The outward-extending tubular projection, E, is particularly troublesome. This type of detail is moldable, but it forms an undercut in the mold. This undercut necessitates the use of a three-piece mold. The third part adds to the cost of the mold, and requires additional labor during the molding process.

There are very few limitations on the location of rotationally molded holes. Where possible, they should be positioned to be in line with the opening of the mold.

The distance between two holes, or between a hole and the outside edge of a part, should be a minimum of three and preferably five times the wall thickness of the part (Fig. 3.29). Less space restricts the flow of the powdered material and produces abnormally thick walls. These thick walls shrink more than the rest of the part and produce molded-in residual stress around the holes. Excess

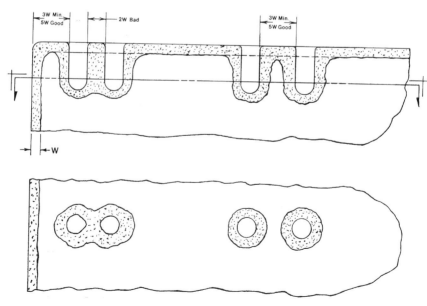

Figure 3.29 Closely spaced hole considerations

shrinkage in the material between two closely spaced holes can pull the material away from the core pins to produce oval-shaped holes.

Some parts require holes for attaching the product to other structures, such as pallets, truck beds, or the floor. This feature can sometimes be provided with a perforated flange on the bottom of the part (Fig. 3.30). The inner column, hole A, is formed over a core pin passing completely through the hollow flange.

The mounting hole, B, is actually an open slot molded into the hollow flange. Two opposing slots of this shape provide a simple but secure mounting method.

Both flange-mounting holes (Fig. 3.30A and B) are in line with the opening of the mold. They do not cause molding or tool-building problems.

The large hole (Fig. 3.30C) at the top of the tank could be provided by machining, as described in Section 3.4.2. An alternative approach that saves

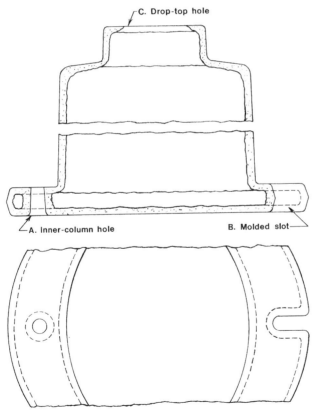

Figure 3.30 Inline tank-mounting holes A and B are easy to produce. An insulated section in the mold can produce a large drop-top opening C

plastic material and sometimes labor is the *drop-out* or *drop-top* technique. With this procedure, the portion of the cavity where a large hole is required is insulated from the heat of the oven. This surface of the cavity may also be coated with slippery, baked-on fluorocarbon or silicone. The theory is that the plastic material will not coat the slippery surface, which takes a long time to reach molding temperature. While using this technique, it is not possible to guarantee a clean edge on the hole. Some of the plastic will extend into the insulated area. Holes of this type normally require some light trimming.

In the case of the open-ended tank (Fig. 3.19), the drop-top technique could be used to create the top opening. An alternative approach would be to use the insulating technique to reduce the thickness or the amount of material in the top portion of the tank to be machined away and discarded or reprocessed.

It is common practice to machine holes into a molded part, but it is generally agreed that they should be molded-in wherever it is practical to do so. Molded-in holes can reduce a part's cost by eliminating secondary operations. Molded holes are generally stronger than machined holes. Once the hole detail is located in a mold, its size and position are fixed. Machined holes are subject to the variations of the machining operations.

Side-acting core pins increase the cost of a mold. Pulling side-acting core pins before demolding the part can cost more than machining the holes as a secondary operation. The location of the core pins relative to other features in the mold may disrupt the flow of the powder through the cavity. Holes that have to be positioned too close to a side wall or another hole should be machined. Some parts are required with and without holes. In other instances, the size or location of the holes changes for different applications.

Drilling is the most common technique used for machining round holes (Fig. 3.31).

Drill presses are used for their inherent accuracy and repeatability while drilling holes in small parts. Handheld electric and pneumatic drills are used for large parts that are difficult to handle.

The location of the hole on the part can be determined by measuring and marking in the standard way. If the location of the hole is constant, a drill-locating dedent can be molded in the part. A trick of the trade is to add the dedent detail to the mold after samples have been molded and the exact shrinkage has been determined.

If the location of a hole is critical, it can be established with a drill fixture (Fig. 3.32). Drill fixtures of this type are frequently used when several holes have to be drilled in precise relationship to each other. The hinge-plate mounting holes in the refuse container lid (Fig. 3.1) are located with a drilling fixture of this type.

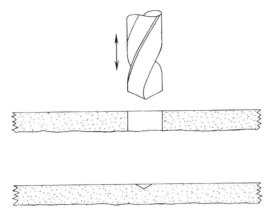

Figure 3.31 Drilled hole located with molded-in dedent

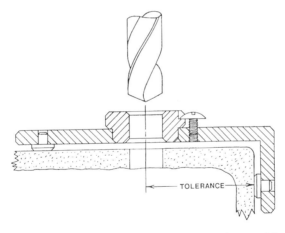

Figure 3.32 Hole drilled with a locating fixture for precision location

Drilling a hole, A, through a wall is the simplest way to provide an opening into a molded part (Fig. 3.33A). The location of that hole and its strength can be improved by molding an inward-projecting closed hole that can be opened by drilling (Fig. 3.33B). The molded recess also has the advantage of positioning the cut edge of the hole away from the appearance surface of the part. The molded recess allows a radius, chamfer, counter-bore, or texture to be molded into the outside appearance surface of the part. A reinforced drilled hole of this type would be worth considering for the heavily loaded, drilled holes that support the wheels on the refuse container (Fig. 3.1). Molded holes of this type would, however, require the use of side-acting core pins.

A partially molded hole (Fig. 3.33C) could be further reinforced by combining an outward-projecting boss and an inward-projecting hole.

Drilling is an efficient and reliable method for machining small round holes. For holes larger than 12.7 mm (0.500 in.), hole saws will produce a cleaner and more accurate opening (Fig. 3.34). Molded-in locating dedents and inward-projecting blind holes are also produced with hole saws. Hole saws are commercially available in sizes up to 152.4 mm (6.000 in.).

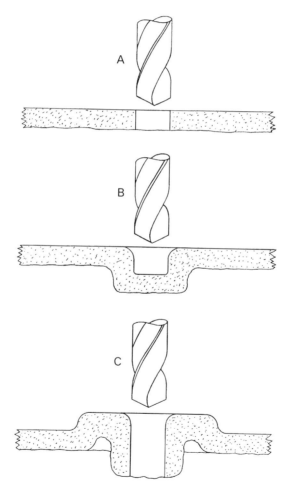

Figure 3.33 A) Simple drilled hole. B) Molded hole locates and strengthens drilled hold. C) Raised boss and molded blind hold opened by drilling to produce strengthened area around hole

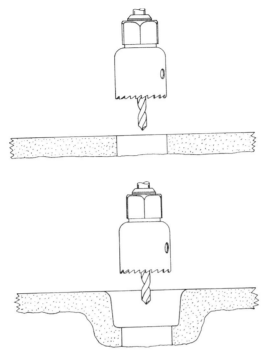

Figure 3.34 Hole saw being used to machine plain and reinforced flanged holes

Holes that are larger than 152.4 mm (6.000 in.), or that are not round, can be machined with saber saws or routers. Routers are most frequently used. Handheld and guided routers with or without a fixture are used for large parts. Small, easy-to-handle parts can be machined with rigidly-mounted routers and guiding fixtures.

Drilling produces round holes with straight walls on the machined surfaces. Routing provides the opportunity to shape the machined surfaces. Chamfers, radii, and counter-bores can be machined into the outside edge of a hole by contouring the router cutter. O-ring recesses have been produced in thick-walled parts by this technique.

The PVC plastisol materials allow holes of any size or shape to be produced by the *tear-out* technique. The eye, mouth, and neck openings on the doll's head (Fig. 3.35) are produced by molding closed holes that can be opened by tearing out a thin section in the wall of the part. The thin wall required for the tear-out is produced with a thin and sharp edge in the cavity (Fig. 3.36). The technique operates on the principle that the thin edges of the metal core on the mold do not have enough heat capacity to produce a wall that is as thick as the nominal wall

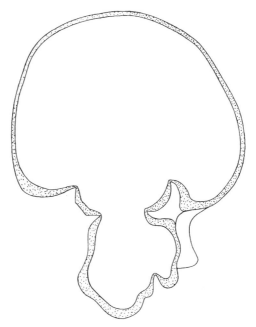

Figure 3.35 Plastisol doll's head with tear-out holes at the neck, mouth, and eyes

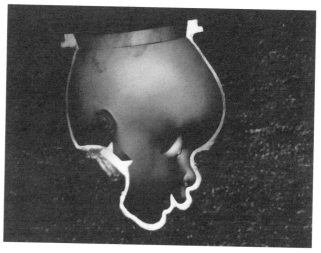

Figure 3.36 Electroformed doll's-head cavity with sharp-edged tear-out surfaces at the neck, mouth, and eyes

of the part. The sharp edge concentrates the tearing force in the thinnest wall to produce a consistent tear-out location.

Head-to-head molding is another technique for producing openings in large parts. A head-to-head molding can be as simple as producing one long part that is cut in the center to produce two open-topped containers. Greater complexity can be achieved by molding two different parts at the same time. The refuse container body and lid (Fig. 3.26) can be molded as one part. The separate lid and the opening into the container are produced by machining a slot between the lid and the container.

3.4.8 Molded Threads

Molded threads appear to be simple shapes, but they are actually quite complex structures. Threads share many of the elements of holes, undercuts, reinforcing ribs, and closely spaced parallel walls. A properly designed thread will comply with the design guidelines for all of these other elements. Inside and outside threads are normally incorporated into molded holes, and are controlled by the kind of holes that can be produced.

Inside threads are undercuts that complicate mold making and molding. Except in very soft materials, inside threads are formed on threaded cores that must be rotated to unscrew the undercut threads. The inner-column hole (Fig. 3.30A) could be threaded by molding the part over a threaded core pin. In that case, the core pin would have to be rotated in order to free the threads. This takes longer than simply pulling a nonthreaded core pin.

Outward-projecting outside threads can sometimes be located on the parting line of a mold (Fig. 3.25, PL 2). In this case, the threads are freed from the cavity in the normal mold-opening sequence. In this location, parting-line mismatch and flash may disrupt the smooth fit between two mating threads. With time, wear at the parting line can cause a thread of this type to become oval in shape. The best quality inside and outside threads are produced with unscrewing threaded cores and cavity inserts.

Threads are actually closely spaced parallel walls. The ideal thread profile would be a shape that provides adequate space for the flow of powdered plastic material (Fig. 3.11). This desirable thread shape is normally set aside in order to provide standard thread forms that are interchangeable with preexisting threaded components. The most frequently specified threads are sharp-pipe and machine-screw threads. In order to be interchangeable with standard thread profiles, these threads are designed to be similar to solid ribs (Fig. 3.15).

Fine-pitched, sharp threads (Fig. 3.37) do not provide adequate space for the powdered plastic material to flow into and coat the thread cavities. A better shape for a rotationally molded thread is the wider profile provided by the Acme or modified-buttress threads. These deeper threads provide increased thread engagement. This allows the threads to perform without maintaining precise dimensions. The increased thread thickness has greater resistance to the shearing off of the thread. Even with these wider thread profiles, the process tried to produce a thick, solid wall. Threads of this type are frequently not completely filled out, and they may exhibit external voids (Fig. 3.38) and sink marks opposite the thread.

In those instances where fine-pitched machine or pipe threads cannot be provided, consideration should be given to machining the threads into or onto molded holes or tubular projections.

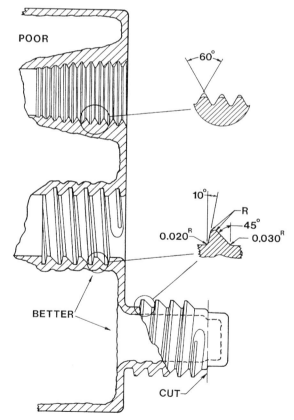

Figure 3.37 The standard 60° sharp threads are difficult to mold. The modified buttress-thread profile is a better shape for rotational molding

Figure 3.38 Incomplete molded thread with large surface voids

In some cases, threads are provided by welding commercially available standard injection molded threaded fittings onto rotationally molded parts as a secondary operation.

3.4.9 Molded-in Inserts

Another approach to providing heavy, load-bearing, small-diameter, fine-pitched threads is with molded-in inserts. Rotational molding is an excellent process for molding in both large and small inserts. Inserts made in metal, plastic, rubber, and wood have been successfully molded. The only limitation in selecting the material for an insert is that the two materials must be chemically compatible. The insert must be able to withstand the heat of the oven. Ideally, the two materials should have similar coefficients of thermal expansion.

Inserts are loaded into the cavity along with the plastic material under room-temperature conditions. During the molding process, the mold, the plastic material, and the insert expand as they are heated in the oven. During the cooling part of the cycle, the mold and the insert contract in size and the plastic material shrinks. Plastic materials expand and contract more than all the metals, but brass and aluminum have the highest coefficient of thermal expansion and are the best metal inserts to use with plastic.

Rotational molding, and transfer and compression molding are the only other processes that heat and expand the metal inserts during the molding process. Other processes mold the hot plastic material directly onto a cool, unexpanded insert. This heating of the inserts has allowed the rotational molding of parts with inserts of 1 m (39.37 in.) or more in length. In the other processes, where the metal inserts are not heated and expanded or cooled and contracted during molding, the differences in contraction and shrinkage prevent the plastic material from shrinking the normal amount. Restricting the shrinkage of the plastic material can result in stress-cracking of the molded part. Problems associated with the differences in thermal expansion can also be encountered in products long after they have been molded. A product that is used outdoors will expand in the summer's heat and contract in winter's freezing temperatures. The larger the insert the greater the differences in thermal expansion and contraction between the two materials will be.

Even with small parts, the plastic material shrinks around metal inserts and is prevented from shrinking the normal amount. This condition creates molded-in residual stress in the plastic material around the metal insert. These stresses will be concentrated on any sharp corners or edges on the insert. The best results are achieved with no sharp edges.

Some molded-in threaded inserts (Fig. 3.39) contain sharp corners that can become stress concentrators. If the plastic material does not coat the insert uniformly, stress cracks can occur.

The plastic materials that have high mold-shrinkage factors are more of a problem in this regard than low shrinkage factor materials. PC has the lowest mold-shrinkage factor of any of the commonly molded materials. Unfortunately, PC is a notch-sensitive material that is susceptible to stress-cracking as it shrinks onto the sharp corners or inserts.

There is no actual chemical bond created between the plastic material and a metal, rubber, or wooden insert. These materials are simply encased by the plastic. Shrinkage of the plastic material onto the insert forms a mechanical bond.

A plastic insert made of the same plastic material being molded can form a chemical bond to a molded part. In order for this to happen, the plastic insert must be heated to its melting temperature. A strong mechanical bond can be

Figure 3.39 Molded-in inserts with sharp corners create molded-in stress

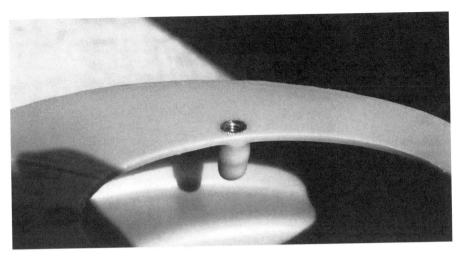

Figure 3.40 Heating the inserts assures that they will be coated with plastic to produce a secure fit

created by melting or welding the two materials together during the molding process. Under the best of conditions, it is difficult to produce a reliable air- or water-tight welding together of the two materials.

The best results are achieved when a plastic or metal insert is heated to the point that the plastic material being molded will adhere to and coat the insert, the same way as it coats the cavity (Fig. 3.40). The design of the insert and the mold must be coordinated to provide for heating the insert up to molding temperature at approximately the same rate as the rest of the cavity. If this is not done, there will be thin walls and weakness around the insert.

Provisions must also be made for securely mounting the insert in the cavity. In this regard, the low pressures and the lack of flow of the melt during the molding process are distinct advantages over the melt-flow processes that tend to unseat or deform inserts. A competent molder and tool builder will know all the techniques for mounting and heating inserts.

There are many commercially available sources for small, threaded inserts (Fig. 3.41). The majority of these inserts provide small grooves and/or knurls that form undercuts for the plastic material to flow into. It is difficult for the powdered plastic material to completely coat these small undercuts. The shallow depth of these undercuts provides only a limited amount of resistance to torque and pull-out forces.

A more reliable and stronger fit will be achieved between the plastic material and the insert if the insert provides large undercuts for the plastic to flow into

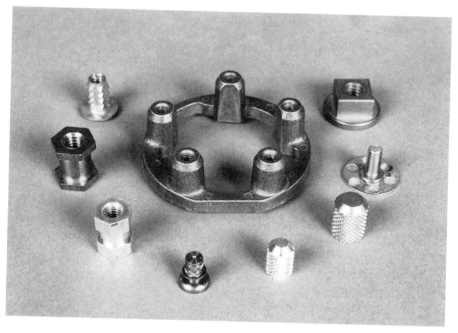

Figure 3.41 An array of commercially available metal inserts. The deeply undercut hex inserts are ideal for the rotational molding process

(Fig. 3.41). Special inserts of this type can be purchased from a few sources, or they can be efficiently produced by conventional screw-machining procedures. Purchasing inserts from a commercial source is almost always more economical than setting up to produce small quantities of special inserts. Standard inserts should be specified where possible.

Molded-in inserts can be located anywhere on a molded part. The efficiency of the molding process will be enhanced if the inserts are located on surfaces where they can be pulled away from their mounting devices as the molded part is removed from the cavity. Inserts positioned with their length parallel to the mold's parting line can be specified, but they must be mechanically released from the mold before the part can be demolded.

Inserts located too close to the side wall of a part can create increases in wall thickness around the insert (Fig. 3.42). The resulting thicker wall will encourage excessive material shrinkage, molded-in residual stress, voids, and sink marks. The distance between the outside surface of a part and the closest surface on an insert must be three, and preferably five, times the nominal wall thickness of the part in order to avoid these problems.

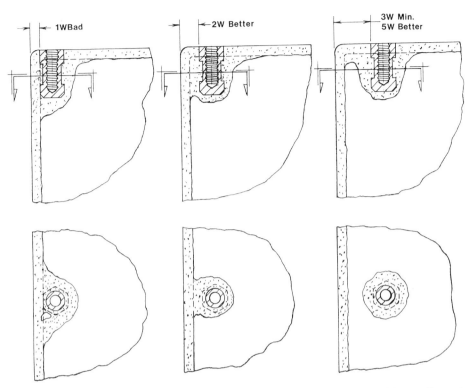

Figure 3.42 Molded-in inserts located too close to side walls create thicker walls and voids

Molded-in inserts have the ability to distribute applied loads over a larger area on a plastic part. The proper use of inserts can produce extremely strong load-bearing parts. All the weight on the deck and the forces developed by the sail on the catamaran boat (Fig. 3.43) are transmitted onto the solid and foamed PE hulls through molded-in metal inserts. It is doubtful that PE would be strong enough for this application without the load-distributing capability provided by the metal inserts.

3.4.10 Tolerancing

International trade has resulted in an increase in the number of companies competing for the same markets. As these multinational companies strive for market share, they have propagated the concept of low-cost, high-quality

Figure 3.43 The center section of this catamaran is attached to the hulls through large molded-in inserts (Photo per Murray Sill, Courtesy Hobie Cat Co., Mfg. by American Rotational Molding Group, Inc.)

products. During the past decade, the producers of durable products have made unprecedented improvements in quality and manufacturing efficiency. Quality is an all-inclusive concept that covers all aspects of a product. The current trend is for *total quality management*, which covers everything from product design, color consistency, neatness of the parting-line scar, and part dimensions. Continuous quality improvement is a worthwhile objective. But there is a practical limit to what can and cannot be achieved within the combined confines of high quality at a low cost. The dimensional tolerances that are commercially acceptable for a molded plastic part are a compromise between cost and quality.

Rotational molding, twin-sheet thermoforming, and blow molding are all open-molding processes. None of these processes can be expected to provide the dimensional tolerances that are possible with closed-molding processes. These hollow-part processes are not capable of controlling the wall thickness or inside dimensions to the tolerances expected with closed-molding techniques such as injection and compression molding.

The tolerances that can be achieved on any molded plastic part are controlled by a large number of interrelated variables. These variables can be subdivided into four categories: the design of the part, the material being molded, the mold design and construction, and the molding of the part.

- Part design: The production of a dimensionally stable, rotationally molded plastic part starts with the design of the part. Errors in judgment in part design cannot be overcome by the skill of a molder or the selection of an ideal plastic material. A small, thin-walled part will always be more dimensionally stable than a large, thick-walled part molded in the same material. The size of the handles on the refuse container (Fig. 3.1) will always be more consistent than the height of the container.

  In specifying tolerances on a product such as the refuse container, the designer should avoid specifying tight tolerances on the long dimensions that control the height and width of the part. This container does not have to fit with any other product. All that is important is that it provides a given volume. A part of this type could be designed with liberal dimensional tolerances, and the product could be controlled by specifying a minimum volume or capacity.

  Closer tolerances could be specified on the smaller handles and their supports if there was a need to do so.

  Rotationally molded parts are produced in molds with no internal cores. Dimensions and tolerances can only be specified on those surfaces that are molded in contact with the cavity. The body portion of the refuse container is free to shrink and pull away from the cavity during the cooling part of the molding cycle. Larger tolerances will be required on the dimensions for the body of this container.

  The pickup truck hook recess (Fig. 3.1) on the refuse container has to fit the hook. These inside dimensions on the part are molded over a core in the cavity. The core prevents the plastic material in this area from shrinking. This allows closer tolerances to be specified on the hook recess dimensions.

  A properly designed rotationally molded part will have different tolerances on different details of the part. These tolerances will reflect what is and is not possible to achieve with this open-molding process. The designer must always resist the temptation to specify one blanket tolerance that applies to all dimensions of a part. Applying the same close tolerances required for the pickup truck hook recess to the whole part would result in an unnecessary increase in part cost.

  The thickness of the nominal wall of a rotationally molded part cannot be specified or controlled the way it can be with closed-molding processes. This topic is discussed in detail in Section 3.4.1 but, in general, a commercial wall thickness tolerance is ±20%. A tolerance of ±10% can sometimes be achieved when wall thickness uniformity is more important than cost.

Details such as closely spaced parallel walls, sharp corners, reinforcing ribs, holes, and inserts that accumulate extra material and create thicker walls should be carefully considered. These thicker walls stay hot longer and shrink more than the surrounding thinner walls. These variations in shrinkage make it more difficult to maintain dimensional repeatability.

- Material consideration: The uniformity of the dimensions of a molded part depend on the uniformity of the plastic material being molded. There are lot-to-lot variations in the plastic material received from suppliers. Mold shrinkage is the most important variable as far as molded part tolerances are concerned. The recent trend toward improved quality has resulted in a significant improvement in the uniformity of the plastic material being received. The increased use of reprocessed materials, off-spec materials, fillers, and additives results in greater variation than would be expected with different batches of premium grades of material.

    Fillers, slip additives, and pigments can all affect the mold-shrinkage factor of a plastic material. For example, it is common practice to mold preproduction samples in uncolored, natural material. These parts may be within drawing specification, but that does not guarantee that the first production run of pigmented samples will be the same size. Preproduction samples that will be used for approval to start production must always be molded with all the additives that will be used in production.

    Close tolerances are easier to achieve with low mold shrinkage factor materials. Crystalline plastics shrink more than amorphous materials. The mold shrinkage of amorphous materials will be more uniform over all areas on a part.

Mold Design and Construction:

- The size of a rotationally molded part is established by the size of the mold's cavity. The size of the cavity is dictated by the part drawing, plus a calculated allowance for shrinkage of the plastic material. If the cavity is not the correct size, there is no way that the part can be molded within drawing specification.

    Theoretically, there is no limit to the level of precision to which a cavity can be built. There is, however, a price to be paid for a higher level of precision. The best mold for any application is the lowest cost mold that will provide the functional and manufacturing requirements of the product being produced. It is not prudent to build a mold to a higher level of precision than is required for the application.

Each mold-building technique has its own level of precision. If multiple-cavity molds are required, the cavity-to-cavity variations must be allowed for. It is generally agreed that computer numerically controlled (CNC) machining will produce the most dimensionally accurate cavity. Depending on the size and shape of the required cavity, CNC machining may or may not be practical.

A good quality precision mold will provide for accurate alignment of the various parts of the mold at the parting line. The mold-clamping method must ensure that the parting lines can always be completely closed in order to minimize the variation in those dimensions that extend across the parting line.

Of equal importance are the thermal characteristics of the mold. Ideally, all inside surfaces of the cavity should heat up to molding temperature at the same time. These uniform temperature considerations must also include the heating of molded-in inserts, loose cores, unscrewing inserts, and hard-to-heat, deeply recessed areas in the cavity. Uniform heating is difficult to achieve, due to the size and shape of some parts and the requirements of providing a strong, durable mold. In spite of these limitations, every effort must be made to produce a cavity that can be uniformly heated. A mold designed and constructed for uniform heating is also a mold that can be uniformly cooled. Cooling has a greater effect on mold shrinkage than heating, and it must be given equal consideration.

Processing Considerations:

- There are many potential variables in the rotational molding process that can affect the size of the part being produced. As an example, if there is a variation in the amount of plastic material charged into the cavity, the wall thickness will change accordingly. The shrinkage and part dimensions will also vary with a change in wall thickness.

   The speed and ratio of rotation determine the number of times a specific location on the cavity passes through the puddle of plastic material, and the direction in which it enters and exits the puddle. A change in these molding-machine settings can affect the uniformity of the wall thickness of a part. The use, or lack of use, of reverse rotation can change the uniformity of the nominal wall thickness.

Many parts are run as family molds on a large, standard grid, along with other similar or dissimilar parts. The molding-machine speed, ratio of rotation, oven temperature, and other processing parameters must accommodate all the parts being molded. The machine settings in these cases are always a compromise between the most and least demanding part being molded.

It is generally agreed that the highest level of precision and repeatability of part dimensions can be achieved by devoting a molding machining, or one arm on a carousel machine, to running a single mold. This approach increases the molding cost, but it does allow for molding parts with closer tolerances.

Variations in oven time, temperature, and air velocity can affect the final part size. The hotter the plastic material becomes, the more it expands. The material will then contract or shrink more as it returns to room temperature.

The speed with which a plastic material is cooled will affect shrinkage. Cooling the material quickly results in a low shrinkage factor. Cooling the material slowly increases shrinkage, but the shrinkage will be more uniform.

A product produced in a mold that does not cool uniformly will have different amounts of shrinkage in different locations on the part. These variations in shrinkage encourage warpage and make it difficult to maintain uniform dimensions.

Dimensions that cross the parting line of the mold will change if the mold is not closed to the same location each cycle. Powdered plastic spilled on the mold's flanges at the parting line can prevent the mold from being closed completely.

Variations in the amount of mold release used can increase or decrease the tendency of a hollow part to pull away from the cavity as the part cools and shrinks. This variable can be minimized by pressurizing the part during the cooling portion of the cycle. The internal pressure holds the part in contact with the cavity as it cools, which produces more uniform dimensions.

The best approach to molding parts to close tolerances is to establish the optimum molding cycle and then maintain those conditions. The most common cause of changes in a part's dimensions are intentional or inadvertent changes in the molding conditions.

The refuse container (Fig. 3.1) is molded as a head-to-head part. The lid and body of the container are affected by the same day-to-day variations in plastic material and molding conditions. In spite of these variations, the dimensional relationship of the lid and the container are unaffected, as they both experience the same variations. This is an advantage of the head-to-head molding technique.

There are many other potential variables in the molding process. However, the quality of the rotational molding machines now being produced has minimized these variables. Modern molding machines provide precise control of all aspects of the process. These new controls are accurate and reliable. This has allowed the production of rotationally molded parts with a level of precision that was not possible in the past.

It is worth noting that among the four variables that affect dimensions, two are more variable than the others. Once the part design is finalized and a mold is built, these two elements are no longer variables. That leaves the plastic material and the molding process as the two remaining variables. Between these two, the molding process provides the most opportunities for variation. This is normally the best place to start looking for the causes of unexplained changes in part dimensions.

In spite of all the opportunities for variation that can affect dimensions, the rotational molding industry has established guidelines for dimensional

Table 3.5 **Industry-Standard Dimensional Tolerances in ±cm/cm and ±in./in.**

Plastic material		A	B	C	D*	E	F
PE	Ideal	0.020	0.020	0.020	0.020	0.015	0.010
	Commercial	0.010	0.010	0.010	0.010	0.008	0.008
	Precision	0.005	0.005	0.005	0.005	0.004	0.004
PP	Ideal	0.020	0.020	0.020	0.020	0.015	0.010
	Commercial	0.010	0.010	0.010	0.010	0.008	0.008
	Precision	0.005	0.005	0.005	0.020	0.004	0.004
PVC	Ideal	0.025	0.025	0.025	0.025	0.015	0.015
	Commercial	0.020	0.020	0.020	0.020	0.010	0.010
	Precision	0.010	0.010	0.010	0.010	0.005	0.005
Nylon	Ideal	0.010	0.010	0.010	0.010	0.008	0.008
	Commercial	0.006	0.006	0.006	0.006	0.005	0.005
	Precision	0.004	0.004	0.004	0.004	0.003	0.003
PC	Ideal	0.008	0.008	0.008	0.008	0.005	0.005
	Commerical	0.005	0.005	0.005	0.005	0.003	0.003
	Precision	0.003	0.003	0.003	0.003	0.002	0.002

Note: This table refers to Fig. 3.44
Note: Does not include cavity tolerance
* Plus 0.250 cm for parting-line variation

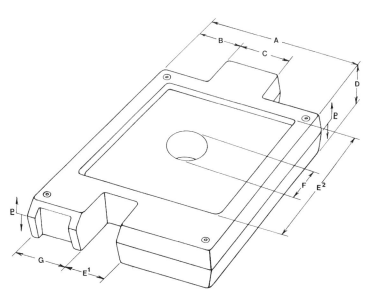

Figure 3.44 This figure is used in conjunction with the list of acceptable tolerances in Table 3.5

tolerances (Table 3.5). The values are in cm/cm and in./in. for the different kinds of dimensions of a hypothetical rotationally molded part (Fig. 3.44). Different values are indicated for the five commonly molded materials. It is important to note that there are major differences in the tolerances that are practical to expect while molding the different materials.

Note also that some locations on a part can be held to closer tolerances than others. For example, dimensions A, B, C, and D (Fig. 3.44) are outside dimensions that are free to pull away from the cavity as the part cools and shrinks. These dimensions cannot be held to as close a tolerance as the inside dimensions E1, E2, and especially F. As the plastic material in these locations cools, its shrinkage is limited by the cavity. These dimensions can be held to closer tolerances.

The four molded-in metal inserts are also held in place as the plastic material around them shrinks. Their location can have the same tolerance as dimension E2.

Outside dimensions C and G are both free to pull away from the cavity as the material cools and shrinks. Dimension G can be held to a tighter tolerance than dimension C because of the recess in the wall in that area. The core that forms the recess also minimizes shrinkage of the outside walls in this location. To a lesser extent, the core that forms the recess at dimension E2 will help stabilize dimension A.

Dimension D crosses the parting line of the mold. This dimension requires the indicated tolerances, plus 0.25 mm (0.010 in.), to allow for variations in the closing of the mold's parting line.

The tolerance values listed in Table 3.5 have been further classified by three levels of precision:

- Ideal tolerances, which can be achieved with a minimum of care. All other things being equal, a part designed according to these ideal tolerances will be the lowest in cost.
- Commercial tolerances, which are possible with nothing more than reasonable care. Any good quality, experienced molder should be able to maintain commercial tolerances.
- Precision tolerances, which require special care and attention. These tolerances should only be specified when maintaining a dimension is more important than part cost.

There is another classification of tolerances that is beyond the industry's recognized guidelines. These are the *fine* tolerances that can only be specified after consultation with a molder.

The tolerance guidelines presented here will be attainable in most situations, but there are always exceptions. There are instances where the material being specified, combined with the size and shape of a part, make it impossible to maintain these recommended tolerances.

The critical nature of some dimensions justifies the machining to size of the dimension after molding. Machining adds to a part's cost, and thus should only be considered as a last resort.

In the final analysis, the best tolerance for a molded part is the largest tolerance that will satisfy the functional requirements of the product. There are no prizes for overspecifying tolerances. There are, however, cost penalties for demanding a higher level of precision than is required.

4   Rotational Molding Molds

Molds are an indispensable part of the rotational molding process, but no one wants to invest the time and money required to build a mold. What customers want are molded parts, and they view a mold as a necessary evil.

Molds have many functions to perform, but their primary purpose is to define the shape of the required part. As engineers design a plastic part, they are also designing the cavity of the mold. The product's chances of success will be enhanced if the designer is familiar with mold design and construction. This chapter is intended to provide some insight into the requirements of the different types of molds used by the rotational molding process.

4.1   Mold Nomenclature

Each new mold represents a significant investment in time and money. It is important to build the best mold for the job. It is equally important to do the job right the first time. The purchaser of a new mold can improve the chance of success by discussing the project with a knowledgeable molder and, preferably, the mold maker before placing an order.

Rotational molding molds are composed of many interrelated parts. Each of these parts has a name. Molders and mold makers refer to the different parts of a mold by the industry's jargon. It is helpful to be able to understand this terminology, in order to avoid miscommunicating while discussing a mold with a supplier.

A typical mold might be a fully framed, cast-aluminum cavity with flat flanges and toggle-clamp mold-closing latches (Fig. 4.1). The entire assembly, with all its attached pieces, is referred to as *the mold*.

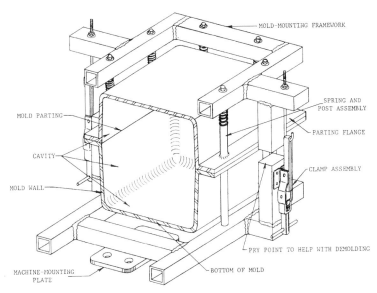

Figure 4.1 Common industry terms used to identify the various parts of a rotational molding mold

The *cavity* is actually only the open space in the mold that forms the part. It is not uncommon for someone to use the word *cavity* to identify not only the cavity per se, but the whole casting that contains the cavity.

The cavity of a rotational molding mold is heated and cooled during each cycle. The walls of the cavity are kept thin in order to expedite heating and cooling. These thin-walled cavities are called *shell molds* or *shell cavities*.

The *mold parting* is where the two or more parts of the mold fit together to create the cavity. The location of the parting line is important, as it will leave a visible mark or scar on the molded part. The parting-line location is also important for efficient molding and for maintaining the mold through its useful lifetime.

Since the walls of the cavity are relatively thin, they are reinforced with *flanges* at the parting line. The flanges provide space for alignment features that locate the different parts of the cavity relative to each other.

Flanges also provide a convenient location for mounting the *posts*, or *pillars*, that are used to attach the cavity to the rest of the mold (Fig. 4.2). Molds may or may not be fully *framed*. Some molds have only a partial frame on the bottom part of the mold. These partial frames provide a method for mounting the mold

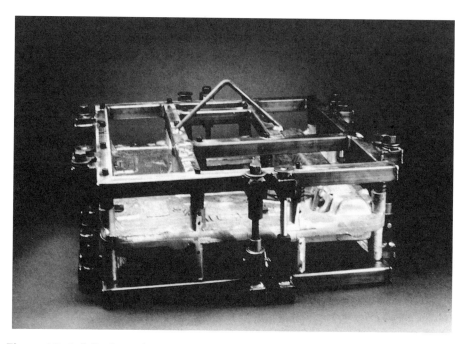

Figure 4.2 A fully framed, cast-aluminum mold with bolt clamping, a positive mold-closing stop, and a mold-lifting attachment (Courtesy Kelch Corp., Cedarburg, WI)

onto the molding machine. A molding machine *mounting plate* is welded to the bottom of the frame. It is important to know what kind of molding machine will be used to mold the parts so that the correct hole pattern can be drilled into the mounting plate.

Frames are normally made of square steel tubing, which provides a high stiffness-to-weight ratio. Frames can be made of stainless steel in instances where long life or corrosion are of concern.

Mold frames add to the cost of a mold, but it is generally agreed that a good quality mold will have a full frame. If a mold is dropped during shipping or molding, the frame protects the relatively thin and fragile cavity. A framed mold also provides something besides the delicate parting line to set the mold on when the mold is opened for demolding.

With a fully framed mold, the parting-line *clamping mechanisms*, such as toggle clamps or bolts, can be attached to the frame and not the cavity. Clamping force is then transmitted from the frame through *springs* to the posts, which actually apply force to the mold's flanges and parting lines. Some molds are built without post springs, but this requires precision clamping to guarantee that the parting lines remain closed. The springs apply uniform pressure along the length of the parting line as the cavity and frame expand and contract as they are heated and cooled.

Frames, springs, and posts are most often used to apply clamping force to the flanges of a mold, but they can also be used to support large, open areas on the cavity casting. This additional reinforcement is important on large molds, or while running pressurized cavities.

A fully framed mold allows the two halves of the frame to be equipped with *pry-point* pads that can be used to open the mold. Pry points of this type discourage machine operators from forcing screwdrivers or crowbars between the parting-line flanges to pry a mold open.

There are different kinds of molds, but the individual parts of the different types share the same nomenclature.

4.2   Types of Molds

The cavity is the most important part of a mold. There are four primary methods for producing cavities for the rotational molding process. Each method produces a different type of cavity. Each type has its own unique capabilities and limitations. Selecting the optimum type of cavity for a given product is another important part of the product development process.

While considering the different types of available mold-building techniques, it is worth remembering that the actual molding of the plastic material in the cavity is accomplished with little or no pressure. This is a low-pressure process. However, the handling and clamping of the mold and the demolding of parts require that the mold be strong. Building strong, thick-walled cavities and frames adds to the mass of the mold that must be heated and cooled during each cycle. The weight of the cavity, the frame with all its accessories, plus the weight of the plastic material, must be supported by the arm of the molding machine. These considerations always result in a compromise between cost, strength, weight, and efficient heating and cooling. There is no one mold that is ideal for all applications. The different molds reviewed here are listed according to their frequency of use. The capabilities and limitations of each type account for why they were chosen, and that is a good guide for anyone who has to choose a mold for a new project.

4.2.1 Cast Aluminum

Aluminum's good thermal conductivity, light weight, and ability to be cast into intricate, complex shapes makes it the most frequently used cavity material for the rotational molding process. Cast-aluminum cavities are specified for small to reasonably big parts that will be produced in modest to large quantities.

Aluminum cavities are cast in low-cost sand or plaster molds, using specialized foundry techniques. The sand and plaster molds are formed over patterns that duplicate the shape of the required part (Fig. 4.3). The patterns can be made of any rigid material, but hardwood is the most common.

Pattern makers are highly skilled craftspeople who are guided by their customers' part drawings. The resulting pattern will be a replica of the part to be produced, plus an increase in size to allow for shrinkage of the cast aluminum (1.1%) and the plastic material (0.5 to 3.5%). Cast cavities cannot be better than the pattern, which can only be as good as the part drawing. Omissions or errors in judgment in designing the part will be duplicated in the pattern, the cavity, and the molded parts.

Figure 4.3 Original pattern and the two cavity halves cast from that pattern (Courtesy Kelch Corp., Cedarburg, WI)

One advantage of cast cavities is that the pattern can serve as a preproduction prototype. By studying the pattern, a designer has a chance to check the dimensions and to see what the three-dimensional part will look like or how it will feel. If changes are required, they can be made before time and money are lost casting the cavities and finishing the mold.

It is not uncommon for a pattern maker to discover that drawing dimensions do not add up, or that multiple angles do not come together as envisioned. This is not the time to be making design changes; however, patterns are relatively easy to modify. It takes time and costs money to change a pattern, but only a fraction of what would be involved if errors were not detected until the mold was built and sampled.

One of the disadvantages of cast cavities is that it takes time and money to build a good quality pattern, and that adds to the cost and delivery of the mold. On the positive side, the pattern allows a final check on the design before the mold is built and sampled. Once a pattern exists, it can be used to quickly produce a second or third mold at a much reduced cost.

The advent of computer-aided design and computer-aided machining (CAD/ CAM) eliminates misinterpretation of part drawings, while improving the accuracy and reducing the delivery and cost of patterns.

In the early days plaster molds produced much better quality castings than sand molds. Sand molds continued to be used due to their lower cost [19]. Usable castings were produced in sand molds, but porosity and the inability to replicate fine surface details limited their use. Over the years the quality of the sand and binders improved. Refinements in casting procedures and the improved sand and binders now allow sand molds to produce better quality castings.

Molders and mold makers have strong preferences for both sand and plaster-cast cavities. Plaster molds cool the aluminum slower than sand molds. Some mold makers believe that slow cooling of the aluminum casting produces a more homogeneous grain structure with uniform physical properties through the thickness of a casting. Plaster molds are still favored for castings containing fine surface detail such as leather graining. It is generally agreed that the differences in the quality of plaster and sand castings continue to be reduced.

Depending on the size of the cavity being cast, sand-cast cavity-wall thicknesses can run all the way from 6.4 to 12.7 mm (0.250 to 0.500 in.). Small cavities, or sections of a cavity, have been cast with thicknesses of only 3.2 mm (0.125 in.). Plaster-cast cavities have thicknesses in the range of 4.7 to 9.5 mm (0.187 to 0.375 in.). The average thickness for both types is 6.4 mm (0.250 in.).

The thickness of a cast cavity is important, as it affects the weight of the mold and the time required to heat and cool the cavity. Within the confines of being strong enough to withstand the rigors and abuse of the molding process, the thinner the cavity wall, the better.

The thickness of the cavity wall is also dictated by the aluminum's ability to flow through the sand or plaster mold. There is no agreement on the maximum practical size of a cast-aluminum cavity. There is, however, a length that the aluminum can flow before it becomes too cool to produce a good quality casting.

The slower cooling rate of plaster molds tends to allow the casting of larger, thinner castings. Both sand and plaster molds have been used to cast cavities as long as kayaks and canoes. Other large cavities have been produced by welding or bolting together two smaller cavities. It is generally agreed that it is difficult to produce good quality castings if their depth exceeds 91 cm (36 in.).

Both sand and aluminum cast cavities have the advantage of allowing parting-line interlocks, mounting lugs, support posts, and bosses to be cast in. The ability to provide reinforced areas by incorporating ribs or varying the wall thickness is another advantage. These special details can be cast onto a mold as one integral part [20]. For example, to provide a threaded neck on a gas tank, an internal thread is machined into an aluminum insert and the cavity is cast with a seat that the insert is fitted into. Another example is a small, heavy boss cast on the cavity that can be machined to hold a core pin to be inserted into the cavity.

Aluminum has the ability to reproduce fine detail, such as leather grain, or to be shot-peened with various sizes of shot and at different pressures to produce special surface finishes. Aluminum cavities can be nickel-plated, anodized, polished, and etched. Polishing and etching can, however, open up subsurface porosity.

Aluminum molds can be welded and repaired easily. Dimensional changes can be made without too much difficulty.

One of aluminum's major limitations is that it is relatively soft, and users of these molds must exercise care during the molding process to avoid damaging these cast cavities.

4.2.2   Fabricated Sheet Metal

Fabricated cavities are second only to cast aluminum in their frequency of use. This type of cavity becomes the logical choice as the size of the required part increases and the complexity of the shape decreases.

Fabricated cavities are built by master craftspeople who employ the techniques developed over many years by the tin-smithing trade. These

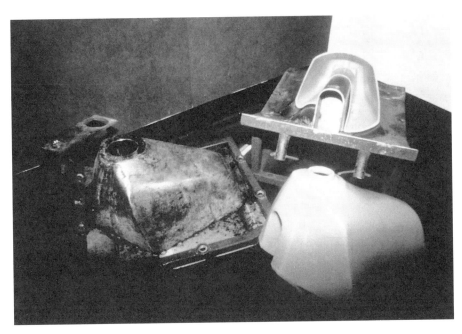

Figure 4.4 Fabricated stainless steel mold and the PE off-the-road bike tank produced by that mold

processes involve cutting, bending, stretching, forming, welding, and machining sheet-metal plates. Cavities of surprisingly complex shape can be produced by these procedures (Fig. 4.4).

Fabricated cavities start out as flat plates that are converted into the required shape (Fig. 4.5) [21]. The accuracy of this part of the work will be improved in the near future with the computer software programs that are now being developed for the sewn fabric industry.

Some shapes that are easy to produce by casting increase the difficulty and cost of the fabrication process. For example, two flat plates can be welded together at a corner. This produces a sharp corner that is not a good shape for molding. Providing a radius at the corner of a fabricated mold requires extra work. It is a simple task to shape a pattern to include a radius that increases in size along the length of a cast cavity. A nonconstant radius is difficult to produce by the fabrication process. Draft angles can turn a simple five-sided box-shaped cavity into a complex combination of angles that have to be carefully fitted together.

The most commonly used materials are mild steel, aluminum, and stainless steel. The thicknesses of these cavities are in the range of 1.3 to 3.6mm (0.050

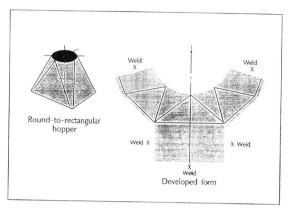

Figure 4.5 A fabricated mold layout showing the progression from a two-dimensional sheet-metal plate to a more complex three-dimensional cavity (Courtesy T.J. Technical Services, England)

to 0.141 in.) for stainless and mild steel. Aluminum cavities have wall thicknesses in the range of 2.0 to 6.4 mm (0.078 to 0.250 in.). The thickness of fabricated cavities is determined by the size of the part, the strength requirements, and the desire to provide a lightweight mold that can be quickly heated and cooled.

One major advantage of fabricated cavities is that different thicknesses of sheet metal can easily be welded into a cavity where extra strength, or more or less heating, is desirable. Aluminum and copper sections have been incorporated into steel cavities to increase thermal conductivity in hard-to-heat and cool locations.

Cast and fabricated cavities compete with each other for the same applications. It is generally agreed that fabricated molds can be built more quickly and at a lower cost than cast cavities. But there is overlap between the two types. For example, the cavity for a 208-liter (55-gallon) barrel could be cast or fabricated. If one cavity was required, fabrication would have a cost and delivery advantage. If multiple cavities were needed, casting would be a better choice. On the other hand, the free-formed contours and surface-appearance requirements of a small-volume bumper-car body would favor one cast cavity. In the case of multiple-cavity molds, the cavity-to-cavity dimensional variations would be minimized by the casting process.

Fabricated cavities are at their best with large parts with relatively simple, smoothly blended contours. Cast cavities excel in producing molds with undulating parting lines. A flat parting line in a single plane is preferred for

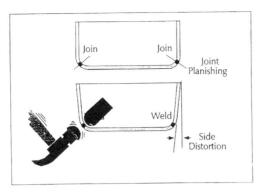

Figure 4.6 The welded joints in sheet-metal fabricated molds require additional finishing considerations (Courtesy T.J. Technical Services, England)

fabricated cavities. Fabricated cavities are built without a pattern, and they do not suffer from the same size limitations as cast cavities. A tank mold with a diameter of 3.7 m (12 ft.) and a length of 9.1 m (30 ft.) has been made by fabrication.

Mild steel and aluminum are the most frequently used materials for fabricated cavities. Stainless steel is more difficult to weld and to form into small radii. Stainless steel is the hardest of the three materials and it will provide the highest polish. Stainless steel resists corrosion, which helps preserve highly polished surfaces. The poor thermal conductivity and weight of stainless steel are limitations. Stainless steel is specified for its strength, corrosion resistance, and ease of maintenance, in spite of an approximate 30% increase in cost.

All the different kinds of fabricated cavities can be polished, blasted, and etched. However, the welded joints present problems. These welded areas can be peened (Fig. 4.6), ground down, and polished, but this can be a labor-intensive and costly undertaking [21]. It is difficult to avoid pits and porosity in the welds. These cavities can be chemically etched, but the welded joints etch differently from the nonwelded areas.

4.2.3 Electroformed

When the rotational molding of plastic made its debut in 1946, the electroformed tooling industry was already well developed. Some of the first products produced in PVC were dolls' heads, balls, figurines, and artificial fruit molded in electroformed cavities.

The most commonly electroformed materials are copper, nickel, and combinations of the two. Some of these cavities are chrome-plated to improve their durability and corrosion resistance. Nickel cavities have a wall thickness of approximately 2.0 mm (0.080 in.). Copper cavities are usually about 3.2 mm (0.125 in.) in thickness. Electroforming is capable of producing wall thicknesses of 6.4 mm (0.250 in.) on large cavities requiring additional strength.

The maximum size of an electroformed cavity is limited by the size of the electroforming tank. A number of mold makers have 2.4 m (8 ft.) long tanks. There is reported to be at least one tank in Japan that is 6.1 m (20 ft.) in diameter. As a general rule, electroformed cavities are limited to a length of 2.4 m (8 ft.). There is no minimum size limitation.

Electroformed cavities are considered to be some of the highest quality and the highest cost molds being used for the rotational molding process. These cavities are specified for applications requiring the replication of extremely fine details, and for the production of multiple-cavity molds requiring exactly the same dimensions in each cavity. Electroformed and cast cavities are specified for geometric shapes that would be difficult or impossible to make by machining, or to form with sheet-metal plates. Electroformed cavities find wide use for deeply undercut parts that have to be produced in hidden parting-line cavities. An example of a hidden parting-line molded part could be the head of a mannequin that is collapsed by a vacuum and demolded through a small opening located at the base of the mannequin's neck. In this case, there would be no visible parting line on the appearance surfaces of the head.

Electroformed cavities are formed by electroplating onto a removable mandrel or pattern (Fig. 4.7). These molds have the same advantages and disadvantages as those associated with the patterns required for cast-aluminum molds. Deeply undercut and hidden parting-line cavities are formed over mandrels that can be collapsed, dissolved, or melted out of the cavity. Decorative figurines and dolls'-head cavities are sometimes formed over the artist's original carved wax master. Other mandrels are made of low-temperature melting metal alloys or foundry sand-filled vinyl skins.

Electroformed cavities have the advantage of being relatively thin and lightweight. Copper has the best thermal conductivity of any of the materials used for building cavities. Copper's thermal conductivity is eighteen times faster than stainless steel's.

Nickel and copper cavities can be polished to a high finish, but subsurface porosity can be a problem. There are definite limitations on the polisher's ability to reach the whole surface on deeply undercut electroformed cavities. These cavities can be blasted to produce special finishes on a part. Chemical etching is done, but this can open up subsurface porosity. In most cases, it is better to

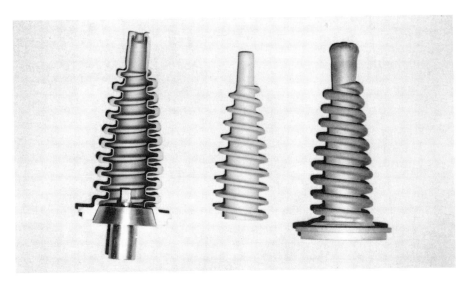

Figure 4.7 A bellows mandrel and an electroformed nickel cavity. Note the increased thickness of nickel on the outside edges of the sectioned cavity

polish or etch the mandrel, and rely upon the process to duplicate that finish in the cavity.

4.2.4   Machined

Mild steel, aluminum, and stainless steel can be machined to produce rotational molding cavities. Machined cavities have always been an option, but their usage in the past has been limited. Machining is used for small, simple shapes that have to be put into production quickly or that require a level of precision that cannot be achieved by the other cavity-making processes (Fig. 4.8).

The thin-walled shell molds that are ideal for this process require machining not only the cavity, but also the outside of the cavity. Some shapes do not lend themselves to the easy removal of metal from the outside surface of the cavity. Some cavity details, such as undercuts, cannot be reached by a cutter, so these cavities have to be made in several pieces. Support posts and other special features normally have to be made separately and assembled to the cavity. Machining also wastes a lot of material.

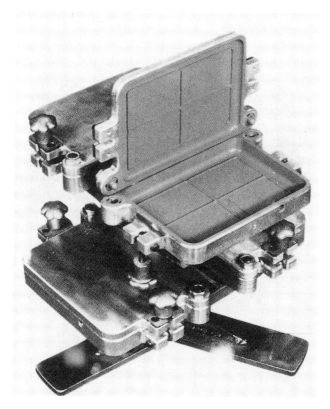

Figure 4.8 Machined, four-cavity, aluminum, color-plaque mold

The thickness of machined cavities can be whatever is required for strength and efficient heating and cooling.

In spite of these limitations, machined cavities have an impressive list of advantages. Machining can provide the shortest mold delivery. Patterns are not required, and mold construction can start as soon as a purchase order is received. There is some time in developing the cutter-path program, but once this is done for the first cavity, the second, third, and fourth cavities can be built in an even shorter time at a lower cost. There is no limit to the level of precision that can be provided on a machined cavity.

Most machined cavities have been made of aluminum, but machining provides the maximum number of choices for selecting a material for the cavity. These machining grades of aluminum are fully densified and they do not contain

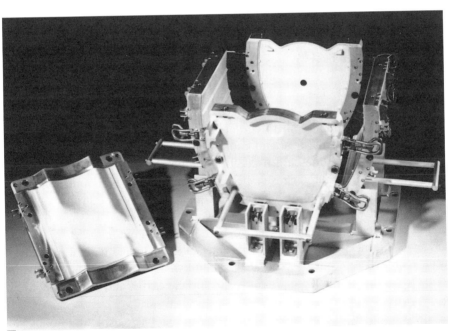

Figure 4.9 Open and closed views of fully CNC-machined, aluminum mold (Courtesy Wheeler Boyce Company, Stow, OH)

porosity. Machined molds allow a higher level of polishing than can be achieved with electroformed or cast cavities. Machined cavities are the best cavity for chemical etching. Engraving, shot-peening, and blasting with various abrasive materials are common cavity-finishing techniques.

Machined cavities are now being reevaluated. As more and more designers shift to three-dimensional solid model databases for drawing, CAD/CAM is becoming a realistic technique for building rotational molding cavities (Fig. 4.9). The ability to go directly from a CAD database into a tool-path program and on into a CNC machining center is now bringing machined cavities into the same cost range as the other mold-making procedures. There is a very high probability that the improved quality and reduced delivery of CNC-machined cavities will result in wider use of this technique in the foreseeable future.

4.2.5   Others

Theoretically, rotational molding molds can be made of any material that will withstand the high oven temperatures encountered in this process. Experience has isolated casting, fabrication, electroforming, and machining as the best processes for building molds, but there are some exceptions.

Vapor-formed nickel cavities have been and are still being used for rotational molding. These cavities are similar to electroformed nickel cavities, except that they are more costly. They can be made in very large sizes. They are known for their excellent high-luster surface finish. The absence of the typical electroforming buildup on sharp corners allows the production of cavities with uniformly thin walls that heat and cool uniformly (Fig. 4.7).

Sprayed-metal cavities are used for prototype parts and short production runs. They are generally chosen because they can be built quickly with a minimum of cost. Sprayed-metal cavities require a pattern. The surface quality is not good. The inherent porosity of sprayed-metal cavities is such that they require heavy coatings of mold-release, both inside and out, in order to fill in the porosity. The metal is brittle and will crack if not handled carefully. Sprayed-metal molds should not be used at temperatures above 316°C (600°F). Water cooling creates too much of a thermal shock to the cavity, and cooling should be by ambient or fan-blown air.

Jacketed molds heat and cool the cavity through passages built into the wall of the cavity. Hot and cold oil can be pumped through these passages to mold plastic parts without the use of ovens or cooling chambers. Casting or machining the passageways and pumping hot oil through rotary unions, are costly, troublesome, and maintenance-prone. This has discouraged the use of jacketed molds. Improved aluminum-casting procedures and the inherent efficiencies of CNC machining are causing this type of mold to be reconsidered. With a few exceptions, the work being done today is of an experimental nature. In the future, this may become an important mold-making and molding option for certain shapes of parts.

Composite molds, made of glass fiber reinforced epoxy, have been used in the past. An alternative approach is a cast-silicone rubber cavity backed up with a glass fiber reinforced epoxy shell or chase. These cavities are low in cost, but they are severely limited by the poor thermal conductivity of the nonmetallic materials. The successful work done to date has been with liquid exothermic thermosetting plastics that internally develop the heat required for curing or cross-linking.

A new development that combines the advantages of cast-composite cavities and jacketed molds is the Wytkin CMT Technology. This technology involves a totally new approach. A cast-composite cavity has electrical heating elements embedded just beneath the molding surface of the cavity. Cooling is achieved by blowing air through cast-in passages. The Wytkin CMT molding machine is reviewed in Chapter 7. It is too early to determine whether this breakthrough mold-making technology will be widely accepted by molders. The opportunities promised by this new development are so impressive that it is bound to find a place in the industry.

4.3   Thermal Management

If a mold's primary purpose is to define the shape of a part, its next most important function is to conduct heat into and out of the plastic material.

All the thermoplastic molding processes heat and cool the plastic material, while the temperature of the mold remains constant. The rotational molding process heats and cools both the mold and the plastic material. Everyone wants cavities built of materials that have high thermal conductivity, in order to minimize the time required for heat to pass through the wall of the cavity. Copper conducts heat faster than aluminum, but copper is weak and easily damaged. Heat will pass through aluminum many times faster than through mild steel, nickel, or stainless steel. In spite of this limitation, cavities are being built using all these metals.

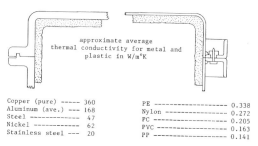

Figure 4.10 All metals conduct heat so much faster than plastic materials that the differences among the mold-building metals lose significance

Rotationally molded parts are produced in thin-walled, shell-type cavities. Mild steel and stainless steel are stronger than copper or aluminum, which allows cavities with thinner walls to be built with these materials. A thin-walled, mild-steel cavity can conduct heat into and out of the plastic material at the same or a faster rate than an aluminum cavity with a thicker wall (Fig. 4.10).

The thickness of a cavity's wall and its rate of thermal conductivity are important but overrated considerations. What is of more concern is the rate at which the plastic material can conduct heat from the cavity through the wall of the plastic part. PE has the highest rate of thermal conductivity of any of the commonly molded plastic materials. The rate at which heat passes through a solid, molded PE wall is only a fraction of the speed with which heat passes through the wall of a stainless steel cavity. A solid PE part will conduct heat five times faster than PE powder. All the commonly used cavity metals conduct heat so much faster than plastic that the differences among the metals lose significance.

The rapid thermal conductivity of all types of cavities actually causes problems. During the heating portion of the cycle, the plastic material closest to the cavity is heated to a higher temperature for a longer time than the material on the inside wall of the molded part. During the cooling portion of the cycle, the outside surface of the part is cooled before the inside surface. This condition results in a difference in shrinkage and crystallinity between the inside and outside portions of the molded wall. The resulting molded-in stress within the wall contributes to warpage and a loss in impact strength and temperature resistance. Once a mold is built, the thermal conductivity of the cavity material and the plastic material cannot be changed. The molding cycle can, however, be controlled to minimize the different rates of heating and cooling within the wall of a molded part. Modern molding machines provide the capability of profiling the oven temperatures based on what is happening inside the cavity. Slow, gentle cooling will produce a more uniform rate of cooling through the wall of the molded part.

In the final analysis, uniform heating and cooling of all the plastic material is more important than fast heating and cooling. The molding conditions can be manipulated to improve uniformity, but the best way to ensure uniform heating and cooling is in the initial design of the mold. An ideal situation would be where all the inside surfaces of a cavity reach molding temperature at exactly the same time.

Building cavities with a uniform wall thickness is the best way to ensure uniform heating and cooling. Even with the best of intentions, it is not always possible to produce a cavity with a uniform wall thickness. A cast-aluminum mold will almost always have an increase in thickness where the two flanges meet at the parting line (Fig. 4.10). The inside surface of the cavity at the parting line will always take longer to heat and cool than the other, thinner portions of the cavity. A fabricated cavity (Fig. 4.10) with welded-on angle iron flanges would allow more uniform heating and cooling of the entire cavity.

Relatively thick-walled flanges and bosses, of the type used to mount core pins, inserts, and clamps, can also distract from uniform heating and cooling of a cavity.

Mold designers and mold makers are in the best position to influence the uniform heating and cooling of a cavity. Mold makers and designers are, however, limited by the design of the part to be produced. If a design engineer draws a double-walled part with a deep recess, it is a foregone conclusion that the outside of the cavity will reach molding temperature sooner than the recessed areas (Fig. 3.7). There are ways to heat these deep recesses, but they all increase mold cost and are never 100% effective.

Closely spaced holes or reinforcing ribs sometimes prevent a mold maker from providing a thin-walled, shell-type cavity between these details. Specifying a solid reinforcing rib (Fig. 3.15) instead of a hollow rib creates a thicker wall. These thicker sections take longer to heat than the other, thinner portions of the part.

If a designer understands the process being used, there are many little things that can be incorporated into the part design that will improve the cost of molding and the quality of the molded part. For example, kiss-off ribbing is a common reinforcing detail on parts with closely spaced parallel walls (Fig. 4.11). The hollow cores that form round tack-offs are miniature recesses that are harder to heat and cool than other, more exposed surfaces of the cavity. Designing a tack-off into both walls of a part reduces the depth of the recess.

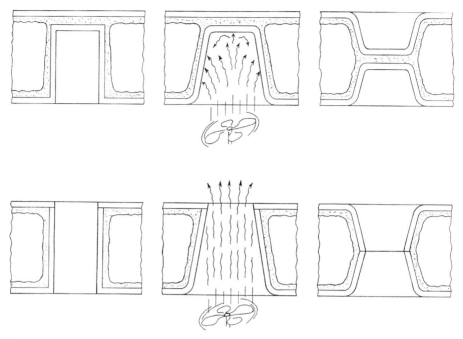

Figure 4.11 The deep, blind recesses required for kiss-offs and tack-offs are more difficult to uniformly heat than shallow, open recesses

Reducing the width-to-depth ratio of tack-offs produces a more uniform cavity temperature and a more uniform wall thickness.

An open tack-off that passes completely through both walls of a part is always easier to heat and cool than a closed tack-off.

There is nothing wrong with designing a part such as a double-walled flower pot (Fig. 3.7) with a hard-to-heat recess, as long as the designer understands the implications of that undesirable feature. On the other hand, if the designer has the option, it would be better to design a wider flower pot with a shallower recess.

4.4   Containing Mold Cost

The molds used by the rotational molding process are low in cost, compared to those required for producing the same size and shape of part by other hollow plastic part processes. In spite of these low tooling costs, no one wants to invest any more than necessary in a mold for a new, untried market.

Molds are built by skilled craftspeople who are in short supply and who are therefore well paid for their services. All mold-making processes are labor-intensive in varying degrees. Both customers and molders blame mold makers for what they consider to be the high cost of molds. In most cases, mold makers are only guilty of giving the molder and the customer what they ask for.

A molded part can only be as good as the mold that produces it. There is no substitute for a good quality mold. Investing a lot of money in a mold does not guarantee that a good quality mold will be built. Mold cost can be managed and molds do not have to cost a lot of money.

The different molds vary in cost. Once a mold type has been chosen, the cost of that mold will be determined primarily by the design of the plastic part. Engineers normally have many design options within the confines of the functional requirements of a product. This allows designers to specify details that will increase or decrease mold cost.

Generally, small molds cost less than large molds. Parts with simple shapes cost less than those with complex shapes. Overriding both size and shape considerations are dimensional tolerances. The tighter the tolerances, the more costly the mold will be. Designers should always strive to specify the broadest tolerance that the application will tolerate, in order to minimize both mold construction and molding costs.

Incomplete part drawings or drawings containing errors are a commonly cited cause for increased mold costs and delivery. Giving a mold maker an incomplete part drawing leaves things for the mold maker to figure out. This is an open invitation to misinterpretation and a loss in both time and money.

The industry is now in a state of transition, but three-dimensional solid model CAD-generated drawings or databases are growing in usage. These CAD drawings allow faster and more thorough communication between design engineers and mold makers. Anything that improves communication is desirable.

While a design engineer is working out the design of a plastic part, the location of the parting line should be established as soon as the overall shape of the part can be determined. Once the parting line is established, the direction of the draft angles can be determined.

The location of the parting line also indicates how the mold will be opened to demold the part. Every effort must be made to design the plastic part in such a manner than it can be produced in a two-piece cavity. Cavities made up of more than two parts cost significantly more due to the added complexity of aligning and clamping the extra parts of the cavity.

Undercuts always complicate mold construction, and they should only be specified as a last resort. If they cannot be eliminated, they should be kept as shallow as possible. The shape of the undercut and the surrounding surface on the part must be designed to assist in the demolding of the part. The indiscriminate use of undercuts can result in a multipiece cavity.

While determining the location of the parting line, the design engineer must allow for enough space for the powdered plastic to be placed in the one part of the cavity that is attached to the molding machine.

Fabricated and machined molds will be easier to build with uniform radii and draft angles. The blending of different radii and draft angles along the length of a part is less of a problem with cast and electroformed cavities that are built using patterns.

The rotational molding process is not good at producing parts with variations in wall thickness. Design engineers should strive to specify uniform wall thickness on parts to be made by this process. Some small, gradual changes in wall thickness are possible, but the required procedures always increase both mold cost and processing difficulties.

Provisions must be made to locate and secure molded-in inserts in the cavity. Keeping the number of inserts to a minimum not only reduces mold cost, it also simplifies the molding of the part.

No part drawing should be released for mold building without a clear definition of the required surface finish. Overspecifying a surface finish can add significantly to the cost of a mold. Polishing a cavity to a high sheen can account for as much as 30 to 50% of the total mold cost. Chemical etching is the second most costly cavity finishing technique. Shot-peening costs less than blasting the cavity with abrasive media. The lowest cost cavity-finishing technique is the ill-defined *industrial finish*. Different mold makers have different interpretations of what constitutes an industrial finish. It normally means that machining tool marks and the scratches left by coarse sanding will be visible on the cavity. An industrial finish should not be specified without a clear understanding of what the finish will look like on the molded part. The ideal finish for a plastic part is the lowest cost finish that will be accepted in the marketplace.

The design engineer's control over the cost of a new mold is second only to the influence of the molder. Experienced molders have strong opinions regarding the type of molds they are going to run. Customers are looking for the lowest cost mold. Molders want a reasonably priced mold that will run efficiently for a long time with a minimum amount of problems. Molders are willing to invest money in features that will make the mold easier to run or maintain. A molder will be concerned with such things as uniform heating and cooling of the cavity, how the molded-in inserts are attached to the cavity, whether there is room enough in the cavity for the powdered plastic, how the part will be demolded, and perhaps whether there will be special clamping features to aid in quickly attaching the mold to the machine. These important considerations may justify a redesign of the molded part. Design engineers must remain open-minded to the molder's recommendations for design changes and request for a better quality, high-cost mold. The molder's suggestions will normally result in better quality, lower cost parts and a long-running mold.

There is no substitute for a good quality mold. At the same time, the best mold is the lowest cost mold that will satisfy all the requirements of the product. The lower cost mold will, however, not always be the best investment. The building of a new mold is a one-time investment. The molding of the parts is an expenditure that is made hour after hour, day after day and if the product is a success, year after year. Within reason, a more costly mold that allows the production of higher quality, lower cost parts is a better investment.

A new product's chances of success will be enhanced by starting with a plastic part that is designed to accommodate the rotational molding process's unique capabilities and limitations. One of this process's major limitations is that many of the products being molded are designed by engineers with no prior experience with rotational molding. One of the recurring problems in this area is that these inexperienced designers fail to ask for help. Experienced molders, plastic material manufacturers, and mold makers are aware of this lack of prior experience and they can be counted on to help in the finalizing of a part design.

4.5   Selecting the Optimum Mold

Every new plastic product requires a mold that will convert the liquid or powdered plastic material into the shape that is required. The purchase of a new mold is a serious undertaking. This is a new product development function that must be done thoroughly and carefully. This is not a process that can be rushed. Many well-designed plastic products have failed or suffered throughout their existence because shortcuts were taken in the mold-selection and building process.

The selection and construction of the optimum mold for a new product is a complex undertaking that justifies the input of an expert. The purchase of a new mold is not, however, something that the design engineer can afford to ignore. Molds are not standard commodities that can be left to the discretion of a buyer. At this point in the project, the design engineer knows more about the new product than anyone else. Choosing the type of mold to be built is a decision

Figure 4.12 A sheet-metal fabricated mold for a roll-out refuse container mounted on the arm of a rotational molding machine

that should be shared by the design engineer and the suppliers. The experience gained in this close involvement will be invaluable for future projects.

It is difficult to be definitive about how to select the best type of mold for a given application. The roll-out refuse container mold (Fig. 4.12) was built by the sheet-metal fabrication process. That same mold could also have been made by any of the other common mold-making techniques.

It goes without saying that the best mold is the lowest cost mold that will satisfy all the marketplace and manufacturing requirements. At the same time, the molded part can only be as good as the mold that produced it. Each new tooling program, therefore, becomes a compromise between cost, quality, delivery, ease of molding, and mold life. The capabilities and limitations of the different commonly used molds are reviewed in Section 4.2.

Each mold has its own inherent advantages and disadvantages. The capabilities of the various types overlap one another. There are many instances in which the same part could be satisfactorily molded with several different kinds of molds. For example, a cavity for a round ball (Fig. 4.13) could be made by any of the commonly used mold-making techniques. The football, with its

Figure 4.13 The different sizes and shapes of parts now being rotationally molded require different kinds of molds

molded-in leather grain, stitching, lacings, and nonsymmetric shape, would be more difficult to produce by machining or sheet-metal fabrication. The rabbit figurine and the football are of shapes that could best be accommodated by casting or electroforming.

A cavity for the large tank, with its plain surface, could be made by any technique. If only one cavity was required, sheet-metal fabrication would be the lowest cost, fastest way to get into production. If the tank was not too large and more than one cavity was required, casting would have the advantage. Building the tank cavity by machining or electroforming would be possible, but not economically justifiable, but there are always exceptions. If the tolerances on the tank were tight, or if the surface finish was demanding, electroforming or machining might be worth considering.

A customer wants a long-running, low-cost, quick-delivery mold that will produce good quality, low-cost molded parts. Molders want the same thing, except they also want a mold that is easy to run and that doesn't cause any trouble. The wants and needs of both the customer and the molder have to be met in order to have a win-win situation.

Molders have very strong opinions on the type of mold they want to use for a given project. Prior experience and personal preference are a molder's major considerations in deciding which type of mold construction is to be used. Aside from personal preference, cost and delivery are probably the next most important criteria to be considered. Other factors to be taken into account are such things as the shape, size, strength and durability, and appearance requirements of the molded part; thermal management, ease of use, and weight of the mold; and number of molded parts and cavities, level of precision, and degree of detail required.

The best way to ensure a successful mold selection is good planning. Good planning involves a thorough review of the project by the designer and the molder. Working together, they should establish how many parts will be required and how many years the product will be in the marketplace. Do the holes have to be molded-in or could they be drilled? What is an industrial finish? How will the cavity be built? Will the mold be fully framed? What kind of clamping will be used to close the mold? Does the product really require tight tolerances on all dimensions or just on fitment surfaces? Is a Class A finish really required? Both must listen carefully during these discussions, as there is often a difference between what is asked for and what is actually needed. If the product development checklist (Table 2.9) has been filled out, that list and the part drawing will contain most of the information required to make a good mold selection decision. Prior to meeting with a molder, the design engineer can study a copy of the checklist and highlight those items that related to mold making. Reviewing the checklist refreshes the designer on the product's requirements and serves as a good preparation for a meeting with a molder.

The mold type comparison (Table 4.1a,b) can be helpful as a quick reminder of the attributes of the commonly used mold types. This table is generic in nature. It does not represent any one product. The ratings are the author's; best estimates for a hypothetical part that could be molded with all types of molds. The values assigned to each mold type and each attribute might change, depending on the size and shape of the part. For example, the lowest cost mold for one cavity of a large, round tank would be a sheet-metal fabricated mold. The lowest cost mold for a rabbit figurine would be casting. In spite of these obvious limitations, the chart is useful in quickly comparing the attributes of the common mold-making procedures. The numerical values assigned each mold type–attribute combination are indicated with ten being the best or most frequently specified combination. A rating of five or above is worth considering. A value of one indicates that this combination is rarely specified, or that it can only be done with significant difficulty.

Regrettably, the rotational molding industry has acquiesced to its customers' demands and has perpetuated the concept that this process can function successfully with cheap molds. There is a very strong human tendency for both

Table 4.1a   Mold Type Comparison Table

	Cost	Delivery	Prior experience	Pattern required	Multiple cavities required	Large size capability	Shape complexity	Hidden parting line	Fine details	Precision
Cast aluminum plaster	8	7	10	Yes	9	8	9	6	9	7
Cast aluminum sand	9	8	5	Yes	9	9	9	5	8	7
Aluminum fabricated	9	8	8	No	4	10	7	1	6	5
Steel fabricated	10	8	9	No	4	10	6	1	5	5
Stainless steel fabricated	8	8	7	No	4	10	5	1	4	5
Nickel electroformed	6	6	6	Yes	10	7	10	10	10	9
Copper electroformed	7	6	5	Yes	10	8	10	10	10	9
Machined aluminum	5	10	4	No	6	6	4	1	5	10
CNC-machined aluminum	6	9	3	No	8	6	7	1	6	10

Table 4.1b   Mold Type Comparison Table

	Surface finish	Lack of porosity	Weldable	Corrosion resistance	Thickness (mm)	Thickness (in.)	Density (gm/cm³)	Density (lb/in³)	Thermal conductivity (W/m°K)
Cast aluminum plaster	6	8	8	9	4.7–9.5	0.187–0.375	2.63	0.095	156
Cast aluminum sand	6	8	8	9	6.4–12.7	0.250–0.500	2.63	0.095	156
Aluminum fabricated	8	10	9	9	2.0–6.4	0.078–0.250	2.91	0.105	180
Steel fabricated	9	10	10	5	1.3–3.6	0.050–0.141	7.83	0.283	47
Stainless steel fabricated	9	10	7	10	1.3–3.6	0.050–0.141	8.03	0.290	20
Nickel electroformed	7	8	9	9	2.1–6.4	0.081–0.250	9.05	0.327	62
Copper electroformed	7	9	9	8	3.2–6.4	0.125–0.250	8.94	0.323	360
Machined aluminum	10	10	9	9	3.2–9.5	0.125–0.375	2.91	0.105	180
CNC-machined aluminum	10	10	9	9	3.2–9.5	0.125–0.375	2.91	0.105	180

10 = Best or most common   5 = Worth considering   1 = Not normally done

molders and their customers to favor the lowest cost mold-making process. The reduced quality and useful life, and added cost of running a mold from the lowest bidder may offset the low initial cost. There is an old adage that "a penny saved is a penny earned." In the rotational molding industry, a penny saved on a mold is not necessarily a good bargain.

5 Understanding the Process

Rotational molding is a material-based process, but the industry takes its name from the unique machines that heat and cool plastic material in a cavity simultaneously rotating in two directions. Customers use this process to produce hollow plastic parts that they can sell at a profit. Customers rely on a molder to use molding machines to produce acceptable parts. Most customers ignore what goes on in the factory until something goes wrong. By then, it is too late to avoid major problems. A rotationally molded product's . chances of success will be greatly enhanced if the customer's design engineers and buyers have a basic understanding of the process and business practices of this industry. No attempt is made here to teach the reader how to perform rotational molding. Some insight as to what happens on the factory floor will, however, allow customers to use this unusual process to its maximum capabilities.

One of rotational molding's limitations is that watching the process being performed creates the impression that it is a simple, low-tech procedure. Nothing could be further from the truth. Anyone who takes control of one of these molding machines and attempts to mold two identical parts will immediately come to realize that this is a complex process, with many variables that must be controlled. Controlling these variables to produce an acceptable part is the stock in trade of a molder.

In the business of producing plastic parts, it is clearly understood and supported by law that a customer or OEM of durable products is the entity that knows, or should know, the most about the functional requirements of a product in the hands of the end-user. It is the customer's and the design engineer's responsibility to design a plastic part, select a suitable material and process combination, and establish quality specifications that will ensure that the product functions as required.

It is the molder's responsibility to use a molding machine and a mold to produce a part according to the part drawing in the specified material, to the quality levels established, at the price and delivery agreed to.

At some point in time, all customers have stumbled through their first rotational molding project. Even today, there are many customers who are using this process for the first time. First-time users should know how to design a functionally acceptable product, but what plastic materials to use or how to design a part for the rotational molding process may be beyond their expertise. Experienced molders, material manufacturers, mold makers, and consultants can be helpful in providing that needed expertise. All of these suppliers are wonderful resources for help in developing a new product, but they will probably not know as much about the functional requirements of the product as the customer does. The best approach is for inexperienced customers to consult with and consider their suppliers' recommendations. It must be remembered, however, that it is the customer who will ultimately introduce the product into the stream of commerce and assume the responsibility for its suitability for the intended purpose.

5.1   Types of Molding Machines

The molding-machine manufacturers have been creative in developing different machines that perform the basic elements of the rotational molding process. Most customers are not concerned with the type of machine being used. In the final analysis, all these molding machines do the same thing: they heat and cool a biaxially rotating mold. The machines are all capable of producing the same hollow parts, but there is a difference in the way they mold them. Some insight into the advantages and disadvantages of each machine can be helpful in using this process to its maximum capabilities.

There are four elements of the process (Fig. 5.1).

A. A plastic material is charged into a mold mounted on the arm of a rotational molding machine.
B. The machine moves the mold into the heating chamber and the arm rotates the mold.
C. At the end of the heating cycle the machine moves the rotating mold into the cooling chamber.
D. Following cooling, the machine moves the mold to the open station, where the molded part is demolded. The process can then be repeated.

All four basic functions can be achieved with one mold mounted on one arm. With a single-arm molding machine, the functions are performed

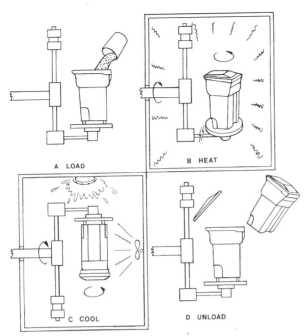

Figure 5.1 A schematic of a single-arm rotational molding machine depicting the four sequential parts of the molding process

sequentially. For a given number of molds per arm, a single-arm machine's output in parts per day is limited.

One method for increasing the output of a molding machine is to perform the various molding functions simultaneously. The need for this desirable method led to the development of the multi-arm turret rotational molding machine (Fig. 5.2). A three-arm machine of this type allows one mold to be heating, while a second mold is cooling and a third mold is being loaded and unloaded. The multi-arm machines cost more than single-arm machines, but their output can be greater.

This more efficient method of molding requires an investment in three molds instead of one. If the volume of the product being produced does not justify the cost of three molds, the same efficiency improvements can be achieved by equipping each arm with a mold for a different product.

Many multi-arm turret machines are run with one relatively large mold on each arm. An alternative approach is to mount several smaller molds on a universal round grid or a specially built *spider* on each arm. All the molds on

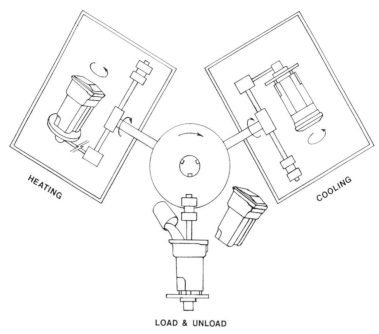

Figure 5.2 A schematic of a three-arm carousel rotational molding machine depicting the simultaneous performance of the four parts of the molding process

one arm could be for the same large-volume part, or they could be for several different smaller volume parts.

The different types of molding machines evolved to satisfy a marketplace need or opportunity. Each type of machine has its own attributes. The single-arm machines may be of the *shuttling* or *swing-arm* type. The swing-arm machines are actually single-arm turret machines that simply swing the mold back and forth between the heating and cooling chambers. The shuttle machines (Fig. 5.3) move the mold lineally into and out of the heating and cooling chambers.

The *clamshell* single-station, single-arm machines are unique in that heating and cooling are accomplished in the same chamber (Fig. 5.4). The clamshell and some of the larger shuttle machines have their arms supported at both ends. The clamshell machines require a smaller space than the equivalently sized shuttle and swing-arm machines.

All the single-arm machines are relatively low in cost for a given size capacity. All three are available in small sizes that are suitable for schools, prototyping, or research and development projects. The single-arm machines are not limited to running a single mold, but are ideally suited to running special

Figure 5.3 A single-arm shuttle rotational molding machine and mold (Courtesy Ferry Industries, Inc., Stow, OH)

Figure 5.4 A single-arm, single-station, clamshell rotational molding machine (Courtesy FSP Machinery, Winkler, Manitoba, Canada)

parts with extreme quality requirements or unusual cycling conditions that cannot be accommodated on multi-arm machines running a mix of different parts. The primary disadvantage of the single-arm machines is that they run sequentially, and their output can be limited for a given number of molds, but there are exceptions.

One misconception in the industry is that the single-arm machines are only useful for low-volume production. Nothing could be further from the truth. Single-arm machines can be equipped with universal grids, or spiders, that will accept a multiplicity of cavities. One single-arm clamshell machine with eight cavities is now producing Chrysler consoles at the rate of 384 parts every eight-hour shift. In other instances one machine operator is able to service two or three single-arm machines with a production rate that is better than what could be achieved with a three-arm turret machine.

For example, three molds could be mounted on a single-arm machine that ran with a fifteen-minute heating cycle, a ten-minute cooling cycle, and a five-minute mold-servicing cycle. That machine would produce three molded parts every thirty minutes and the cycle time would be ten minutes per part.

If those same three molds were mounted one each on the three arms of a small turret machine, the production rate would be controlled by the longest part of the cycle. With the same fifteen-minute heating cycle, that machine would produce three molded parts every forty-five minutes and the cycle time would be fifteen minutes per part.

Rock-and-roll machines are actually specialized single-arm machines. These unique machines rotate, or roll, a mold 360° in one direction, while tipping or rocking the mold approximately 45° above and below horizontal in the other direction (Fig. 5.5). Open-flame heated rock-and-roll machines were one of the first types of rotational molding machines developed for the molding of large parts. Most modern rock-and-roll machines use a forced, hot-air oven to heat the mold. These machines can run multiple molds, but they are normally used to run one large mold. Rock-and-roll machines have all the same advantages and disadvantages as the other single-arm machines.

The lack of a full 360° of rotation in two directions imposes limitations on the shapes of parts that can be molded. Rock-and-roll is not the ideal process for molding round balls. This process is at its best with parts that have a large length-to-width ratio, such as canoes or pipe. For large parts of this shape, rock-and-roll machines are less costly than the equivalently sized multi-arm carousel machine. Rock-and-roll heating chambers can be smaller in volume than the ovens that have to accommodate 360° of rotation in all directions. These smaller heating chambers can result in significant savings in energy costs.

Figure 5.5 A rock-and-roll rotational molding machine and mold (Courtesy Caccia Engineering S.p.A., Italy)

The most common type of rotational molding machine now in use is the multi-arm turret or carousel (Fig. 5.2). These machines are available with up to six arms, but three or four arms are the usual number. Carousel machines are available in a wide range of sizes.

There are two fundamentally different types of carousel machines now in use: the fixed-arm and independent-arm. The three-arm fixed-turret machine (Fig. 5.6) was the most common type of machine in use until 1997. Independent-arm machines, which were introduced commercially in 1982, now account for the largest volume of new machine purchases.

With a fixed-arm turret machine, all three arms must move together. Normally one arm is in the heating chamber, the second arm is in the cooling chamber, and the third arm is in the load and unload area. This is an ideal situation when each arm is equipped to mold the same part. Problems begin to appear when each arm is equipped with a different mold (Fig. 5.7) or a multiplicity of different molds. When this happens, the largest part with the thickest wall will require a longer heating cycle than the thinner walled parts. This results in the most demanding part imposing its cycling requirements on all the other parts being molded on the same machine. A similar situation can

Figure 5.6 A three-arm, fixed-turret, carousel rotational molding machine and mold (Courtesy Ferry Industries, Inc., Stow, OH)

develop when one arm is equipped with a number of molds that are slow to unload and load, or that require the mounting of a large number of molded-in inserts. In any of these situations, one arm can slow down the other two arms.

Some of the parts will not be molded as efficiently as if all three arms were equipped with identical molds or as if those three molds were being run on a single-arm machine.

One answer to this problem can be found in the use of independent-arm machines (Fig. 5.8). The ability to move the arms independently of one another allows the arms to be sequenced at different times to accommodate an array of different sizes, shapes, and thicknesses of parts. If the need arises, an independent-arm machine can be sequenced to function the same way as a fixed-arm machine.

A four-arm machine could be equipped with two heating chambers that would accommodate two arms and their molds. Some four-arm machines are equipped with two cooling chambers, depending on the needs of the parts being molded. Other machines are configured to provide an expanded loading and unloading area that will accommodate two arms.

The primary advantage of the independent-arm carousel machines is the added flexibility that they provide the molder. Independent-arm machines cost 5

Figure 5.7 A three-arm carousel rotational molding machine with different sizes and numbers of molds on the arms (Courtesy Caccia Engineering S.p.A., Italy)

to 15% more than fixed-arm machines of the same capacity. In a growing number of cases, the sequencing flexibility justifies the added cost.

Most carousel machines move the arms in a horizontal direction. There is another multi-arm machine that moves in a vertical direction. This type of machine is sometimes referred to as an *up-and-over* machine. The heating chamber is positioned at the top of the machine, with the cooling chamber at the bottom. The loading and unloading area is at the front of the machine, between the heating and cooling areas. The vertical machines are small to moderate in size. These machines can be very energy efficient, due to their compact heating chambers. They have the distinct advantage of providing the capabilities of a horizontal carousel multi-arm machine, while occupying a much smaller floor space.

There are two new types of rotational molding machines currently under development. These innovative machines are discussed in Chapter 7.

Customers rarely become involved in deciding what type of machine to purchase. Molders normally select the type that will do the best job of molding the range of products that they produce. In selecting a molder, customers do

Figure 5.8 An independent, four-arm carousel rotational molding machine (Courtesy FSP Machinery, Winkler, Manitoba, Canada)

inadvertently determine the type of machine that will be used to produce their parts. Some molders have more than one type of molding machine, but it will be an unusual situation when one molder has all the different types.

A 1997 survey indicated that 59% of North American molders were using independent-arm carousel molding machines, 56% fixed-arm turrets, 23% clamshells, 22% shuttles, and 17% rock-and-roll [22]. Customers will most often encounter molders who are running multi-arm carousel machines, which is an ideal situation. Special product requirements may, however, justify searching out a molder who has one of the special types of machines.

5.2 Molding Considerations

The different types of rotational molding machines produce molded parts in different ways, but they all perform basically the same functions. With the exception of the internally heated, jacketed molds reviewed in Section 4.2.5, all the machines rotate a mold surrounded by hot air in a heating chamber. Today, most heating chambers are of the gas-fired forced hot air type. The length of the heating portion of the molding cycle is determined by the temperature and velocity of the hot air in the heating chamber. The thermal conductivity of the cavity, its thickness, and the general design of the mold will determine how quickly that heat is conducted through the cavity and into the plastic material. The thermal conductivity of the plastic material being molded, the part's wall thickness, and the design of the part will determine the time required for the plastic material to be heated to the point that a good quality part can be molded.

During the cooling portion of the molding cycle, the mold and the molded part are cooled while rotating in room temperature (ambient) air, forced air, chilled air, a mist of air and water, or a shower of water. In most cases, cooling is achieved with a controlled combination of these options. Water cooling is the fastest and most severe type of cooling. Rapid cooling contributes to warpage. Crystalline plastic materials require slow cooling to provide time for the crystals to form. Rapid chilling is economically desirable, but it can have a negative effect on the physical properties of the molded part.

The third and fourth parts of the molding cycle are the loading and unloading, or servicing of the mold. The overall cycle time can be dictated by the length of time required for heating, cooling, or mold servicing. The most efficient molding cycles are achieved by coordinating and controlling part design, material selection, mold design and construction, and molding. The close cooperation of everyone involved in this process cannot be over-emphasized.

It is difficult to be definitive regarding overall cycle times, as they are dependent on many variables. A properly designed thin-walled part produced in a good quality mold can be molded on a cycle of only seven to eight minutes, but an average cycle would be in the range of ten to thirty minutes, depending on the wall thickness. A large, poorly designed, thick-walled part in a low-cost, "cobbled-up" mold could require a cycle time of thirty to sixty minutes, or even 200 minutes in extreme cases. These seem like long cycles compared to the fraction of a minute cycles that can be achieved with injection molding, but the other advantages of the process compensate for the longer cycles. Rotational molding cycles are comparable to the thermoforming of parts with similar wall thicknesses. Both thermoforming and rotational molding have shorter cycle times than the open molding thermosetting reinforced plastics processes with which they compete.

The lowest possible part cost is a much sought after objective by both molders and especially their customers. Part cost relates directly to cycle times. In an attempt to reduce cycle time, molders will do whatever they can to minimize the heating, cooling, or mold-servicing segments of the process. Reducing cycle time is a laudable objective, but it can be overdone. Within the confines of a given plastic material, type of mold, and wall thickness, there is an optimum heating cycle time. There is some minimum amount of time required for the plastic material to absorb enough heat from the cavity and the hot air inside the cavity. If the plastic is not given time enough to absorb that amount of heat, the quality of the molded part will suffer accordingly. XLPE and PVC materials must reach a minimum threshold temperature in order to cross-link or cure.

Finely ground plastic powder is a poor thermal conductor. In addition, the air between the individual particles of powder becomes entrapped between the molded part and the cavity, as well as within the wall of the part. It takes time for these bubbles to be absorbed or to work their way through the wall to the open interior of the part. These entrapped bubbles affect the appearance and physical properties of a molded part.

Laboratory tests have been developed for determining the degree of cure or cross-linking of PVC and XLPE materials. Proper molding of the other materials is determined by the physical testing of molded parts. Low-temperature impact is the most frequently used test, but tear resistance, tensile strength, and other tests are also used.

In the case of PE, a great deal can be determined by visual inspection. In general terms, a PE part with a pitted outside surface and a lot of bubbles throughout the wall will be an undercured part. A smooth surface and fewer bubbles concentrated near the inside surface indicate a slight undercure. A part with no bubbles is properly cured or overcured.

A powdery or wavy dull inside surface indicates an undercured part.

A PE part with a shiny and slightly yellow inside surface is overcured. A darker yellow color and a sticky inside surface is definitely overcured. Overcuring represents thermal degradation of the plastic material, with a corresponding reduction in physical properties.

The degree of cure is an important quality consideration. In general, as the degree of cure increases, there will be an increase in the density of the molded material. An increase in density results in a stronger part with improved impact strength.

A customer who wants to quickly gauge the quality of molded parts can, with a little experience, learn a great deal by simply looking at the parts. If there is a change in surface appearance, bubble content, or inside color, something has changed in the plastic material or the way it was molded, and further testing may be warranted.

The control and repeatability of modern rotational molding machines took a quantum leap forward with the advent of microprocess controllers. These controllers do an excellent job of controlling the time and temperature variables of the process. Heating chamber temperatures and air velocity, as well as direction and speed of rotation, can now be profiled to suit the part being molded. The built-in memory capability allows a return to the exact cycling conditions that produced the last batch of acceptable parts. Molders now have molding machines that allow excellent control over the mechanical variables in the molding process.

One molding condition that has always been problematic is the exact temperature of the plastic material being molded. It has always been possible to measure and control the temperature of the air in the heating chamber, but that is different from the temperature of the plastic material. The nature of this process, with the plastic material inside a rotating cavity, has always made it difficult to determine the temperature of the plastic in real time. Slip rings have been used to attach thermocouples to a cavity, but this approach has proven to be too maintenance-prone for routine production molding.

Determining the temperature of a rotating mold with noncontact infrared sensors is a new technology now being developed by Remcon Plastics' Dr. Paul Nugent and Ferry Industries. This real-time temperature-sensing technology holds great promise for the future.

The Rotolog diagnostic system [23] provides a real-time means of monitoring and controlling the molding conditions by measuring the temperature inside the cavity during the molding process (Fig. 5.9). The sensing unit is mounted on the arm of the machine, it rotates with the mold and transmits signals to a receiver-computer, which continually scans the data and displays it in a graph form. This breakthrough technology was jointly developed by Prof. Roy Crawford and Dr. Paul Nugent with Rotosystems Ltd. in Northern Ireland and Exxon Chemical in Canada.

Infrared temperature sensors and the Rotolog diagnostic system are poised to eliminate this temperature limitation of the process. These two new technologies and the microprocess controller, coupled with the developments in molds and molding machines, have allowed rotational molding to take its rightful place along with the other high-tech plastics molding processes. The industry is now in a state of transition. Some molders continue to practice a process that is as much an art as it is a science. The more progressive molders are embracing the new technologies and are prospering by molding the more profitable, difficult projects in the more demanding materials.

The speed and ratio of rotation are other important molding considerations. These two conditions are not variables. Once they have been established, they do not change. They may change, however, from one production run to another. A customer does not need to be concerned about these two factors. They are under the molder's control. A change in speed or ratio of rotation can, however, explain why one run of parts has a different wall thickness uniformity than a previous run.

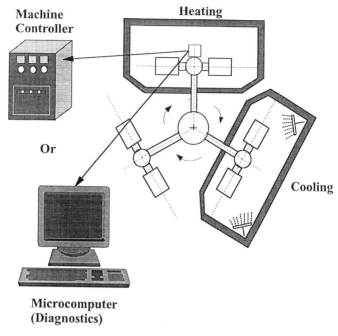

Figure 5.9 A schematic of the Rotolog system where a temperature sensor and transmitter on the arm of a rotational molding machine sends signals to the machine controller or a computer (Courtesy Prof. R. J. Crawford, Queen's University, Belfast, Ireland)

In the case of rotational molding, the speed of rotation of the mold must be slow enough to ensure that gravity holds the plastic material in a puddle in the bottom of the cavity (Fig. 5.10). A high speed of rotation develops a centrifugal force that throws the material to the farthest extremities of the cavity. The cavity for a ball will tolerate a relatively high speed of rotation. A kayak, with its large length-to-width ratio, must be rotated at a slower speed. Rotating a kayak cavity at even a modest speed would develop enough centrifugal force to throw the plastic material to the two extreme ends of the cavity. The resulting part would be thick at both ends, with a thinner wall in the center.

The ratio of rotation refers to the relative number of turns of the major axis and the minor axis. The major axis is the center line of the arm. The minor axis rotates the mold perpendicular to the major axis [24].

The ratio of rotation is determined as

$$\text{Ratio of rotation} = \frac{\text{Major axis rpm}}{\text{Minor axis rpm} - \text{Major axis rpm}}$$

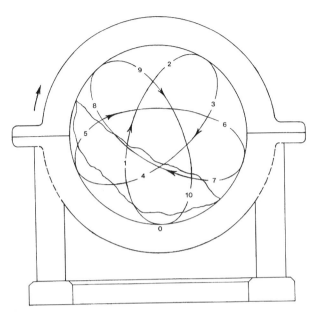

Figure 5.10 A schematic of how one point on a cavity would move relative to the plastic material at a ratio of rotation of four to one

For example, at a major axis of rotation of 8 RPM and a minor axis of rotation of 10 RPM, the ratio of rotation would be four to one, which is a common starting ratio.

The ball mold (Fig. 5.10) illustrates how a single point on the cavity relates to the plastic material as it turns at a four to one ratio of rotation. As the mold rotates, the bottom center starting point 0 moves through the plastic material to location 1 and then on to locations 2, 3, and beyond. Independent of the position of the rotating mold, the plastic remains in the bottom of the cavity. The relationship of the cavity to the plastic material would change with a different ratio of rotation. This relationship would also change if the cavity was the shape of a rectangular tank instead of a round ball.

The molder is responsible for determining the ideal ratio of rotation, based on the size and shape of the part, the number of molds mounted on the arm, and each cavity's distance from the true center of rotation. The final ratio of rotation is usually established, or fine-tuned, by trial and error.

The direction that the cavity takes as it passes through the puddle of plastic material is another consideration. A free-standing detail that projects into the cavity will push its way through the plastic material. This will produce a thicker wall on the leading edge of the projection. Reversing the direction of rotation one or more times during the heating cycle will produce a more uniform wall thickness on the projection.

The relationship of the plastic material to the cavity, as the cavity passes through the material, is important to efficient molding and to the creation of uniform wall thicknesses. No one has ever been able to observe exactly what is happening inside the cavity until Rodney Syler was successful in mounting an insulated video camera and light source inside a mold [25]. This development allowed the detailed study of the relationship between the plastic material and the cavity, and of how the material coats the cavity. In-mold video can be invaluable in establishing the optimum speed, direction, and ratio of rotation.

Customers need to understand what the speed, direction, and ratio of rotation are, but they do not have to be concerned with them, as long as the process is producing acceptable parts.

5.3   Understanding Costs

All OEMs are now under extreme pressure to improve their efficiency in order to remain competitive in the growing global marketplace. Approximately 10% of the plastic products produced in the United States are for the export market. The manufacturers of durable products have become obsessed with controlling and reducing their manufacturing costs. In order to control costs, it has become necessary for the purchaser of plastic parts to develop a better understanding of the factors affecting cost.

In some ways, rotational molding is similar to the other plastics molding processes. In other ways, it is very different. The basic things that contribute to the cost of a rotationally molded part are

1. the design of the part,
2. the plastic material being molded,
3. the design and construction of the mold,
4. the molding of the part,
5. the secondary manufacturing operations, and
6. the packaging requirements.

- Part design: The basic cost of a plastic part is locked in at the time the product is designed and developed. A small, thin-walled, simple part with liberal tolerances can be low in cost. A larger, complex part with thicker walls, tight tolerances, molded-in inserts, and deep undercuts requiring a multi-piece mold will obviously cost more to produce.

- Plastic material: The material specified for molding a part is a significant cost factor. Every effort must be made to use the minimum amount of plastic material. A thin wall with reinforcing ribs can provide the same stiffness with less material than a thicker, nonribbed wall.

- Some PVCs are available as liquids. A few other plastics are supplied as ready-to-mold micropellets. Most plastic materials are manufactured and sold in pellet form. These pellets have to be pulverized into a fine powder for molding. *Toll* pulverizing, shipping, and handling can add 22¢ to 26¢ per kilogram (10¢ to 12¢ per pound) to the material's cost.

- The cost of plastic materials keeps changing, based on inflation, market conditions, and the cost of crude oil and natural gas. The average current costs for the commonly molded materials are listed in Table 5.1. These costs are for large quantities of uncolored, thirty-five mesh powder, except for the liquid PVCs. In addition to the cost of the material itself, there are additional costs for additives such as pigments, ultraviolet light stabilizers, and fire retardants, and for mixing and drying. Some plastic materials and additives require longer molding cycles than others.

- Molds: Some molds are more efficient to operate than others. A two-piece cavity with smooth, drafted surfaces and no undercuts can be an easy mold to service. A multipiece mold with just barely enough volume for the plastic material will slow down loading and unloading of the mold. An abnormally large number of molded-in inserts or several molded-in labels will require more of a machine operator's time.

Table 5.1   Cost of Common Rotational Molding Materials in Large Quantities

	Form	$ per kilogram	($ per pound)
LDPE	Powder	0.27	0.59
LLDPE	Powder	0.26	0.57
HDPE	Powder	0.26	0.57
XLPE	Powder	0.40	0.88
PP	Powder	0.54	1.19
PVC	Liquid	0.34	0.75
Nylon 6	Powder	1.36	2.99
Nylon 11	Powder	2.83	6.23
Nylon 12	Powder	3.18	7.00
PC	Powder	1.22	2.68

- Incorporating quick mold-mounting features, mold-lifting rings, cavity-opening pry point pads, and fast acting release clamps will improve the serviceability of a mold.
- An efficient, good quality mold is rarely the lowest cost mold that can be purchased. Carefully planning a new mold and incorporating efficiency-improving features is normally money well spent. You can pay the mold maker once, or you can pay the molder more for each part. The choice is yours.
- Molding: The actual molding of a part is a major contributor to the total cost of a product. Some molders specialize in one kind or size of parts. Some companies have low overheads and others provide a lot of special services that result in a higher overhead. Some molders are just better than others at managing their companies.

The many different molding machines all have their advantages and disadvantages. Some machines produce parts more efficiently than others. It would be unusual to find a molder who owns all the different types of molding machines and just happens to have open time on those machines. Selecting the ideal molder with the right type of equipment is an important part of controlling part cost.

Running one cavity on a single-arm rotational molding machine (Fig. 5.3) is similar to running a single-cavity mold in an injection-molding machine. Both these machines go through a sequential cycle. A single part is produced and the cycle is repeated. In each case, the one part being produced has to bear the total cost of running the molding machine.

If the same part was molded on one arm of a three-arm molding machine (Fig. 5.2), the parts produced on the other two arms would help pay for the cost of running the molding machine. In effect, three different customers would be sharing the cost of operating the machine. If the three arms of a machine were equipped with standard grids and several different molds, six or ten different customers could be sharing the cost of the machine.

Considering all the possible combinations makes it difficult to understand and estimate the cost of a rotationally molded part. As a starting point, it is necessary to know what type of machine will be used and how many other molds will be run at the same time.

It is also necessary for a customer to understand how a rotational molder defines cycle time. A large, thick-walled part produced with one cavity on a single-arm machine could run on a one-hour cycle. That means that one part is produced every sixty minutes.

Three cavities of that same part could be mounted one each on the three arms of a fixed turret machine. It would still take sixty minutes for any one of the arms to pass through the whole process and produce one part. The cycle required to produce that part is actually sixty minutes, but in that same one hour of molding time, the second and third arms would have produced two additional parts. If the machine produced three parts in one hour, is the cycle time sixty minutes or twenty minutes? Molders refer to cycle time both ways, and a customer must determine which method is being used. The most common method of defining cycle time is as the lapsed time between arms reaching the load and unload station.

Defining cycle time becomes even more confusing with the use of independent four-, five-, or six-arm machines. In these instances, the lapsed-time intervals between arms may not be constant.

A knowledgeable molder will be capable of estimating the cycle for a given part. The actual cycle time may vary from one production run to the next, depending on the type of machine being used and the number of other molds being run on that machine at the same time. There are so many variables involved that a molder may not know the exact cycle time used to produce the required parts on every production run. Customers can avoid this confusion by simply concentrating on the cost of the parts and how many parts can be produced per day. In most instances, it is sufficient to concentrate on the end result and let the molder worry about how to achieve what the customer is asking for.

The other big variable in part cost is the price that different molders charge for the machine being used to mold a customer's parts. Some molders charge different amounts for various sizes or types of machines. Others have one rate for all their machines. Some molders have higher overheads than others. The more sophisticated molders charge some customers more than others based on the kind of additional services that they demand and how much trouble that customer causes them.

Today, there are software programs that allow even small molders to do a thorough job of understanding their costs and arriving at an equitable burden rate. Independent of how the costing is done, all custom molders have to arrive at somewhere near the same shop burden rate in order to remain competitive.

One common method of determining the cost for molding is for the molder to add up the total annual cost of running a factory that does nothing but rotational molding. These costs would include rent, electricity, labor, sales expense, payment to the bank for capital equipment loans, and everything else except the cost of the plastic material, secondary operations, and packaging. This would be the total cost of keeping the machines running. If the molder added to that number the profit he expects to receive for his investment and his efforts, he or she might come up with a number like $3,000,000. If that molder operated six molding machines, each machine would have to generate $500,000 per year to reach the goal of $3,000,000. If each of the molding machines was a three-arm machine, each of those eighteen arms would have to generate $166,666 per year.

Assuming 240 three-shift working days per year, each arm is available for 5,760 hours of molding per year. The catch here is that it is impossible to keep six molding machines running 100% of the time. Machines do break down. Quality control and rejected parts have to be replaced. Additional time is lost in changing and setting up molds, and machines have to be shut down for cleaning and routine maintenance. Production time is lost in running prototypes and sampling new molds. Plastic materials and corrugated boxes are not always delivered on time. Some employees have an annoying habit of not showing up or being late for work.

Another unknown is whether a molder can sell enough work to keep all the machines running full time. Even if this was possible, what happens if one or more of the major customers suddenly loses market share or is burdened with seasonal products? If a molder could keep all the machines running 75% of the time, each of the eighteen arms would only be molding salable products for 4,320 hours per year. Dividing the $166,666 per arm per year goal by 4,320 hours per arm per year indicates that each arm has to generate $38.58 per hour of actual molding.

If this one arm was running one mold on a sixty-minute cycle, the molding cost would be $38.58 per part. If that same arm was running two molds, the molding cost would be half as much. As a further complexity, these two molds could both be producing two head-to-head parts (Fig. 6.6). In that case the cost of the arm would be divided by the four parts being produced. If this arm was equipped with sixteen ball or doll-head molds, the molding cost would be only $2.41 per part.

All molders have their own method of determining cost and burden rate. That burden rate can vary significantly from one shop to the next. Some molders include all labor in their burden rate. Others assign labor based on the number of employees required to service a given mold. The dollar values presented here are purely hypothetical and they are not an indication of actual molding costs. These values are only being used as an example of how a molder with an all-inclusive molding machine burden rate might arrive at a cost using the common method of assigning a burden rate on a per-arm basis.

There are obvious cost advantages to running more than one mold on each arm of a multi-arm machine. An ideal situation is where all the molds on each arm are for the same part. This situation is only possible with larger volume applications. The more common occurrence is where each arm is equipped with molds for several different kinds of products. If one part is a thin-walled ball and a second part is a square tank with a thicker wall, the heating and cooling conditions will be the same for both parts. A third part on the same arm might have a large length-to-width ratio, but speed and ratio of rotation will be the same for all three molds. In this case, the most demanding of the three parts will dictate the cycling conditions of the other two parts. There is a distinct possibility that the thin-walled ball will be overcured or degraded. If this is not the case, the thick-walled tank will be undercured. The long, narrow part could wind up with a nonuniform wall thickness.

If a uniform wall thickness was critical on the long, narrow part, it could be molded under ideal conditions on a single-arm molding machine. In that case, this one part would have to bear the total cost of running that machine. There is a cost penalty for running only one mold on a single-arm machine, but the dimensional tolerances and overall quality that can be achieved are almost limitless.

What a customer normally sees at a rotational molding facility is a state of the art, multi-arm machine with all its cycling conditions automatically controlled by a microprocessor. It seems incongruous, but the molds mounted on that machine will usually be opened and closed, and loaded and unloaded, manually. This looks like, and is, a labor-intensive operation for this day and age. The very thing that holds back automating the mold-servicing part of the process is one of this industry's main advantages, which is the simultaneous molding of different parts on one molding machine. Different molds can be mounted on each arm. One or more arms can be equipped with a standard grid that accommodates several different sizes and shapes of molds. Some of these molds may run for less than a shift until they are replaced with different molds. This scheduling flexibility is important for the efficient production of small quantities and for just-in-time delivery projects.

The standard practice of running a mix of continuously changing molds makes it difficult to automate the mold servicing part of the process. Large-volume applications, such as automobile armrests and

Figure 5.11 An automated, six-arm, fixed-turret, carousel rotational molding machine producing plastisol dolls' heads (Courtesy Ferry Industries, Inc., Stow, OH)

dolls' heads (Fig. 5.11) have been fully mechanized. If the volume being produced justifies it, the process can be automated.

- Secondary operations: Some rotationally molded parts come out of the mold ready to be shipped. Other parts require secondary operations. Some molders are adept at performing secondary operations and solicit that kind of work, while others shy away from it.

Secondary operations can be as simple as pressure-testing a tank for leaks. Other parts may require machining, painting, special decorating, the adding of bar codes, assembly, packaging and, if saddled with a shabby mold, the trimming of parting-line flash. Except for large-volume projects that justify mechanization, these secondary operations are labor-intensive. Secondary operations can be a significant added cost. The best way to minimize secondary operations is to eliminate them. The most rewarding place to look for this type of savings is while the product is being designed. The second best source for savings is in the design of the mold. Paying the one-time cost for side-acting core pins in the mold can eliminate the cost of drilling holes in every part. In other instances the postmold insertion of metal inserts may cost less than molding them in.

- Packaging: Protective packaging is a necessary but unrecoverable added cost. A customer can sell a molded part for a profit, but he cannot sell the package. In fact, there may be a cost associated with disposing of the package. The worst thing about packaging is that it is normally the last thing to be considered. In many cases, the selection of the package to be used is simply left up to the molder. Packaging has to be adequate to protect the product, but it should not be overdone. Where possible, parts should be designed to nest together to minimize packaging. Some parts can be shipped from the molder to the customer in the same box that will be used for shipping the final product. Returnable shipping containers are another possibility. The secret to controlling the cost of packaging is to manage it and not leave it until the last minute. Ideally, protective packaging should be an integral part of the product design and development process.

The current business environment is such that the custom plastic molding industry is under extreme pressure to reduce the cost of molded parts. It is interesting to note that a custom molder is only in control of one of the six cost-contributing factors previously listed.

The most important cost elements are the design of the part and the selection of the plastic material. It is the customer and not the molder who controls these factors.

The quality of the mold that a molder can purchase is limited by the amount the customer is willing to invest in that mold. The extent of the secondary operations is dictated by the design of the product. The packaging is, or should be, specified by and at least approved by the customer. That leaves only the actual molding of the part as a cost factor that the molder can influence.

In most cases, all the custom molders who are invited to quote on a new project will start with the same part drawing, plastic material, quality, and packaging specifications. All molders have the freedom to purchase and use the same type of mold and molding machine to produce the part. The wide spread in cost quotations received from molders is due to the way they manage their companies, the resulting shop burden rate, and how well they use the material, mold, and machine to produce the required parts.

In the final analysis, it is the customer and not the molder who controls most of the factors that contribute to a part's total cost. The single best thing that a customer can do to influence the cost of molding a part is to choose a good molder as a supplier. The ideal molder for the project will not necessarily have the lowest part cost. It is sometimes necessary to spend more in order to maximize profits. The second most important thing that a customer can do to control part cost is to work closely with an experienced molder during the product design and development phase of a new project.

5.4   Selecting a Supplier

The critical function of selecting the optimum supplier or molder for a new project is a complex undertaking. This is too important a job to be left up to a purchasing agent who does not understand the rotational molding business and the process. Too many buyers are only concerned with costs and have no responsibility for the ultimate success of the product. Every customer has the right to negotiate the lowest possible cost. The importance of a low part cost cannot be minimized, but there are many other factors to be considered. Quality, delivery, location, customer service, technical competence, and especially integrity are equally important. These attributes are difficult to reduce to dollars and cents figures, and they are all too often overlooked in making a purchasing decision.

Starting a relationship with a new molder is equivalent to creating a partnership based on the premise that everyone will benefit from the success of a new project. If both partners do not benefit, the partnership will not last and the customer will be forced to find another molder-partner to work with. A good relationship is one of mutual trust and respect, where both partners contribute to and benefit from the success of a new project.

Today, customers purchase parts all over the world with nothing more than three or four faxed interchanges. In spite of the success of this new approach to doing business globally, the old-fashioned face-to-face meeting between a buyer and a seller is still the best. It is never a good idea to turn the implementation of a new product over to an untried supplier without meeting the molder and inspecting the facilities.

A nicely decorated office and well-appointed conference room in a new building filled with state of the art molding machines can be misleading. These visual assessments give an impression of financial success and stability, but they do not indicate whether they have been paid for. What a customer should be watching for is a well-managed company with good quality and manufacturing procedures, carried out by a well-trained, technically competent staff. It is not the high-tech molding machines that are important. What is critical is what the supplier and the staff can do with that equipment. A new microprocessor-controlled molding machine has more capabilities and may be easier to operate than an older machine. A lot of very acceptable products are, however, molded on older machines.

Too many customers are impressed with all the wonderful things suppliers claim they are going to do for them. A better measure of a supplier's capabilities is what has already been done for other customers. The samples from previous projects that are on display in the conference room, are, undoubtedly, the best-looking parts that were ever produced. The parts in the warehouse or on the shipping dock are a better indication of the molder's capability. If all the parts in inventory are dirty and warped, a customer has a pretty good indication that his or her parts will also be dirty and warped.

A molder whose prior experience has been primarily with large industrial parts may not be a good source for small, cosmetically demanding consumer products. A molder with a shop full of relatively small-volume projects may not have the management skills or the machine capacity to handle a large-volume project. It is never a good idea to give a molder a project that requires a lot of secondary operations unless that molder is already performing similar operations. Prime consideration should be given to any molder who is already producing parts that are similar to the new project being considered.

A supplier's list of existing customers is also worth noting. If the list includes General Motors, IBM, and United Airlines, a buyer immediately knows that this molder has had prior experience with high quality standards, cost compression, just-in-time delivery, and dealing with large, bureaucratic corporations. If a molder has only one GM project, it doesn't prove very much. If there are several projects from GM, it indicates that GM believes the performance has been good enough to award the molder more than just the first project. A molder's capabilities do slowly change with time, but the projects that a molder has successfully handled in the past are a very good indication of future capabilities.

Another important consideration is whether a molder has the technical capability or willingness to help the customer's design engineer with part design, suggestions for alternative material choices, and recommendations of the best mold and molding machine. A molder's technical competence and problem-solving capabilities can become valuable later in the project when difficulties arise. This type of customer service is especially valuable to a company that is using rotational molding for the first time.

There are molders who have set themselves apart from the competition by becoming full-service suppliers. These molders provide a full range of services instead of just molding parts. There are exceptions, but providing many peripheral customer services can increase a molder's overhead. In these lean and mean times, many OEMs no longer have a full complement of in-house support staff. In these instances, the added cost of buying from a full-service molder can be a wise choice.

A molder located close to the customer is in the best position to make services available. Many U.S. rotational molders have only one plant, but there is a growing trend for progressive molders to purchase smaller molders. Some molders now have several plants scattered across the U.S. and, in some cases, in other countries. These multiplant molders have a better chance of being close to their customers. Proximity also reduces the cost of shipping large, hollow, lightweight parts. Similar savings have also been achieved by shipping a mold from an east-coast molder to a plant on the west coast.

One of the major mistakes made by inexperienced customers is to purchase a new mold directly from a mold maker. This is never a good practice. A first-time user of rotational molding will not know the advantages and disadvantages of the different types of molds and will probably not be able to thoroughly describe what is needed or be able to answer a mold maker's questions. A novice to this industry cannot be expected to know the fine points of mold design and construction.

Experience has proven that the best chance of success lies in purchasing a new mold through a molder. A knowledgeable molder will have a lot of experience in working with mold makers, and will know the jargon and how the industry works. A molder also knows what is needed in a mold in order to satisfy the customer's needs. The molder will have to work with and guarantee the output from a new mold for the life of the project, and is entitled to the opportunity to influence how a new mold is designed and built.

Some molders will encourage a customer to buy a mold directly from a mold maker. The logic behind this undesirable approach is that it relieves the molder of the cost of financing the mold construction and from any responsibility for a mold that does not perform satisfactorily. A better approach is to make the molder responsible for both the mold and the molding of the parts. The last thing an inexperienced customer needs is to be caught in the middle of an argument between a molder and a mold maker concerning a project that is failing. Accepting the responsibility for a mold and the molding of parts are the stock in trade that molders sell.

OEMs know the most about the functional requirements of their products. A plastic material manufacturer is the expert on materials. Mold makers know more about molds than molders do. The molder is, however, the one supplier who knows enough about all these contributing factors to pull the whole project together and produce an acceptable product. The importance of choosing a good molder cannot be overstressed.

The names of custom rotational molding companies can be found in local yellow pages and plastics industry directories. *Plastics News* publishes its annual ranking of the largest rotational molders in North America in its August issue [22].

The single best rotational molding industry directory is the *Roster*, published annually by ARM [13]. This listing includes all the worldwide molders, mold makers, plastic material manufacturers, consultants, and other suppliers to the industry who are members of the association.

6   When to Choose Rotational Molding

In the past ten years, there has been a noticeable increase in the number of hollow plastic parts entering the marketplace. This increase in usage can be traced to material and processing improvements, coupled with an increased awareness of the advantages of hollow parts by the product design community.

Why should a plastics product designer consider using a one-piece hollow plastic part? Tanks and containers of all types are obvious uses for hollow parts. Rotationally molded dolls' heads, furniture, toys, and machine side panels can be hollow, but they are not tanks. The benefits to be derived from the use of hollow parts are not limited to containers. Once design engineers started taking this fact into account, they found many additional uses for hollow parts.

In the course of designing a new product, a designer will frequently specify two parts to be permanently attached to create a hollow part, such as a planter (Fig. 3.8), or a fuel tank (Fig. 6.1). Whenever this opportunity presents itself, the designer should stop and consider molding these two components as one hollow part.

Product designers have also come to recognize that a hollow part has two walls that can be used to provide two totally different functions. For example, a portable toolchest door with two closely spaced parallel walls could have an attractive front wall with molded-in texture, a graphic, and a handle. The second inner wall could incorporate molded-in recesses for the hinges and the latch. Molded-in holes would provide for convenient mounting of the hardware with thread-forming screws. With an air space between the inner and outer walls, each surface can have a different shape without affecting the other surface. Closed-molding techniques, such as injection, compression, and structural foam molding, cannot provide this degree of design freedom.

The cost of manufacturing materials has increased in recent years. The cost of plastic materials has increased, but not as much as the cost of steel and aluminum. The replacement of metal by plastic is not new, but conversions are now accelerating, with plastic taking on the more demanding load-bearing applications.

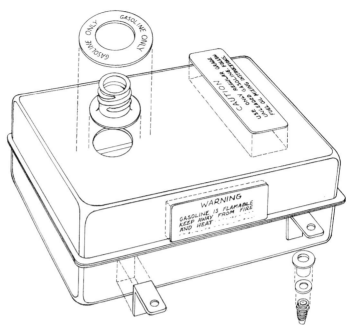

Figure 6.1 A painted, thirteen-piece, fabricated steel fuel tank with seven welded joints and three adhesive-bonded labels

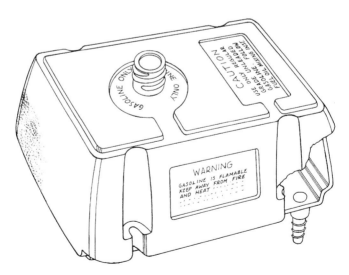

Figure 6.2 A one-piece rotationally molded PE fuel tank that replaced a thirteen-piece metal tank

The average cost of wood increased 36% in North America between 1992 and 1995. This increase now makes plastic economically feasible for products such as furniture, pallets, and construction applications.

OEMs are currently putting a lot of emphasis on parts consolidation. This renewed interest is driven by the desire to increase profits by reducing part count and the number of employees required for assembly. Converting a metal fuel tank to a rotationally molded part can reduce part count and assembly.

Design engineers are familiar with parts consolidation projects, but they normally think of this approach for injection molding. This is an oversight, as rotational molding is also used to combine several parts into one molding. For example, a small fuel tank (Fig. 6.1) was originally made by forming two sheet-metal plates, which were then welded together to form a hollow part. This thirteen-piece assembly required seven welding operations, painting, and the application of three labels. The one-piece rotationally molded replacement tank eliminated welding, assembly, painting, labeling, and the possibility of corrosion and leakage at the welded seams (Fig. 6.2).

Figure 6.3 A rotationally molded, PE farm tractor component that reduces part count by 80% and weight by 65% (Courtesy Rhein-Bonar, Kunststoff-Technik GmbH, Germany)

It is doubtful that this tank could have been produced by blow molding or thermoforming. The degree of stretching required to form the fuel inlet and outlet would exceed the capabilities of these two processes.

The farm tractor component (Fig. 6.3) combines the instrument panel support, molded-in air ducts, floorboards, and two fuel tanks into one part. Part count was reduced by 80%. The part is 65% lighter in weight. It is quieter in use and requires less insulation. It would be inconceivable to attempt to product this component by blow molding or thermoforming.

The floor sweeper frame (Fig. 6.4) is another example of a multifunctional part that eliminates assembly. This one part incorporates all the functions of the 120 metal parts in the previous frame assembly. The two bottom U-shaped recesses fit over the wheel axle and support the total weight of the unit, including a gas engine or an electric motor and batteries. A recess for the filter is molded into the back of the vertical section, which also provides the mounting for the handles. The two front casters are mounted on some of the more than forty molded-in inserts. The hollow housing also incorporates molded-in air-flow passages and the back of the sweeping brush housing. The part is 50% lower in

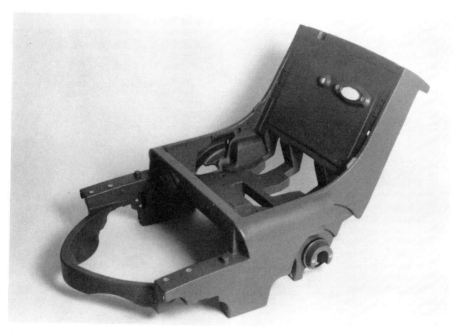

Figure 6.4 A rotationally molded, PE floor sweeper frame that duplicates the functions previously provided by an assembly of 120 metal parts (Courtesy Pawnee Rotational Molding Company, Maple Plain, MN)

weight, which is an advantage for a walk-behind sweeper. The lower weight also reduces the shipping cost and the power required to drive the sweeper. The fully assembled sweeper incorporates ten rotationally molded parts (Fig. 7.3).

The advantages of a hollow part, combined with the ability to replace other more costly materials while reducing part count, continues to open up new markets in industries that have never before considered rotational molding for such technically demanding applications.

The manufacturers of durable products are also placing a premium on early market introduction. The delivery time required for a new mold depends on many factors but, on the average, the molds required for rotational molding can be produced on short delivery schedules.

Relatively speaking, the cost of rotational molding molds is so low that it is not normally economically feasible to build prototype tools. In most cases, the mold used for preproduction sampling and testing becomes the production mold. Once the first samples are approved, there is no additional delay while production molds are being built.

The initial investment required for a new mold is another important consideration. In most cases, blow molding will have the highest mold cost

Figure 6.5 An array of rotationally molded, open-topped PE tote bins (Courtesy Association of Rotational Molders)

among the three hollow parts processes. Rotational molding and twin-sheet thermoforming will have comparable mold costs for simple shapes produced in two-piece molds. For more complex shapes and larger sizes, rotational molding will be the lowest in cost.

The current trend throughout the plastics industry is for larger and more complex parts. The relatively low initial investment required for rotational molding machines and molds gives this process a distinct advantage in the production of large, complex parts that must be produced in small quantities.

Independent of part size and shape, rotational molding is well suited to producing volumes ranging from a few dozen to a few tens of thousands of parts per year. Blow molding is the better process for quantities of hundreds of thousands. Thermoforming is somewhere in between.

Plastics product design engineers think of rotational molding as a process for producing hollow parts, such as large tanks with simple shapes. Large tanks are a major market, but rotational molding is adept at producing small, hollow, and open parts of complex shape. A Barbie doll head is both small in size and complex in shape. This process is also chosen for the production of open-topped drum liners, planters, and all types of tote bins (Fig. 6.5). Many open parts are molded two at a time as one hollow part that is cut apart to produce two parts (Fig. 6.6).

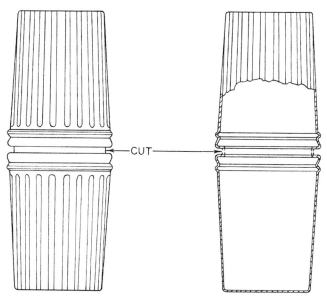

Figure 6.6 Rotationally molded, head-to-head refuse containers. This single molded part is then cut in the center to produce two individual parts

6.1   Competitive Processes

The use of rotational molding to produce open-topped parts brings the process into direct competition with other molding processes such as injection molding, thermoforming, reaction injection molding, dip molding, and injection-molded structural foam, as well as the reinforced thermosetting plastics processes such as hand lay-up, spray-up, resin transfer, and compression molding. In this case, the design engineer has to study all these processes in order to select the optimum manufacturing technique for a new product.

If the design of a new product indicates that a hollow plastic part is required, the design engineer's options are reduced. Hollow reinforced thermosetting plastic parts with high strength and simple geometric shapes can be produced by filament winding, centrifugal casting, resin transfer bag molding, or the lay-up and spray-up processes. These processes all produce high-performance parts that are in a price range well above rotationally molded components.

If the choice of a material-process combination favors a hollow thermo-plastic part, the options are reduced even further. The only mainstream thermoplastic processes that produce hollow one-piece parts are blow molding, rotational molding, and twin-sheet thermoforming

It is beyond the scope of this book to do justice to the important blow-molding and thermoforming industries. A brief review of these processes may, however, be of assistance to the design engineer who has to choose the best thermoplastic hollow-part process for a new product.

6.1.1   Blow Molding

Blow molding is a high-temperature, low-pressure, thermoplastic, open-molding process that uses air pressure to blow a hot preform into the desired shape.

The blow molding process takes many forms. The process of primary interest to the designers of durable products is extrusion blow molding. This part of the industry can be divided into the production of packaging and industrial products. Packaging products are mainly bottles of all types. Small containers and tanks can be either packaging or industrial items, depending on their end-use application. Industrial items include many of the same modest-sized products that are produced by rotational molding.

There are several different kinds of extrusion blow molding machines, but they all perform the same basic function of expanding a hot hollow preform into a larger, more complex shape.

During the extrusion blow molding process, the plastic material is heated in an extruder and not in a heating chamber or oven. The extruder extrudes a hollow, tubular preform downward over a blow pin (Fig. 6.7). The two halves of a mold, which in this case define the shape of a bellows, close around the preform. The top of the hot preform is pinched off and welded closed. The bottom of the preform is clamped onto the blow pin by the closing mold. Air is then injected through the blow pin to expand the preform to the shape of the cavity. While air pressure holds the formed part in contact with the cavity, the part is cooled by water passing through the cooling passages in the mold. After the part has cooled and regained strength enough to retain its shape, the mold can be opened and the molded part removed. The process can then be repeated. Modern blow-molding machines are capable of performing all the molding functions automatically.

A major limitation of blow molding is that it is a stretching process. With a part the shape of a bellows (Fig. 6.7), the preform stretches more to form the crests of the part than to form the root of the bellows. The differences in the

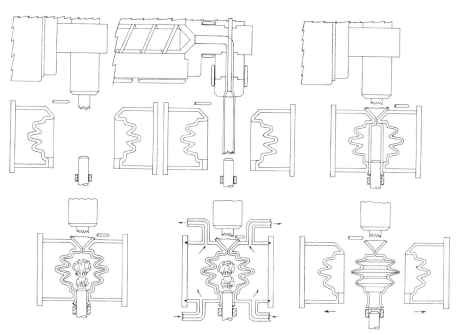

Figure 6.7 A schematic of the interrupted extrusion blow molding process, producing a bellows

amount of stretching of the preform result in parts with nonuniform wall thicknesses. Depending on the shape of the part and its intended use, this thinning of the outside corners of a part may or may not be acceptable.

There are techniques, such as preform programmers, that will produce preforms with thicker walls in those areas that will stretch the most. These preform programmers are helpful, but rotational molding produces parts with more uniform wall thicknesses.

The ideal shape of a part for blow molding is a cylinder that is closed on one end with a small opening at the other end. The best shape for a rotationally molded part is a ball.

Extrusion blow molding machines cost more than rotational molding machines for a given size capacity. Customers and their product designers are not concerned with the cost of molding machines, as they do not intend to buy one. The initial investment in a molding machine does, however, have an effect on a molder's burden rate, or the cost of running the machine.

Blow-molding machines are powered by electricity, which is, on average, 40% more costly than the natural gas that is typically used for heating in the rotational molding process.

The molds required for blow molding are normally cast or machined aluminum. Their cost will be higher than for the equivalent mold for rotational molding. As the size and complexity of a part increase, rotational molding's lower mold cost becomes more obvious. This is especially true for cavities made up of more than two parts.

Blow molding has the advantage of being able to process many thermoplastic materials, including ABS and modified polyphenylene oxide (PPO). ABS and PPO are important materials for the construction, automotive, furniture, electronics, plumbing, and toy industries.

During this process, the plastic material is heated inside the barrel of the extruder. The heating cycle times are short and the hot material is not exposed to the oxygen in the air.

Multimaterial walled parts, especially bottles and fuel tanks, are blow molded, but much more complex and costly molding machines are required.

Some encouraging work has been done with blow-molded foam, but that option is not yet fully developed.

The blow-molding process has the distinct advantage of being able to process materials in their as-received pellet form. This capability eliminates the cost of pulverizing the pellets into a fine powder. Extrusion blow molding produces scrap factors that can be 40 to 50%. The scrap material and rejected parts can, however, be reground on site and fed back into the molding machines with no additional processing.

The cycle times for blow molding are dictated by the time required to cool the part. These parts are cooled only from the side in contact with the cavity. Small, thin-walled bottles have been produced on a ten-second cycle. Some large, industrial parts run on cycles of seven to ten minutes.

Theoretically, there is no maximum limit to the size of blow-molding machine that can be built. There is, however, a practical limit to the size of preform that can be extruded and handled through the process. There is one blow-molding machine in France that is producing 7,570-liter (2,000-gallon) containers. A German company is blow molding 4,996-liter (1,320-gallon) underground storage containers. The cost of these molding machines increases dramatically beyond a 1,135-liter (300-gallon) size capacity.

Two hundred and eight-liter (55-gallon) plastic shipping and storage barrels are capturing an increasing share of the steel barrel market. These barrels are produced by both extrusion blow molding and rotational molding. With a product of this size, shape, and volume, the initial investment in molding machines and molds is approximately the same for both processes. The part costs of the two processes are competitive. Blow molding is preferred for larger volume, lighter duty barrels. Rotational molding dominates the market for smaller volume specialty barrels with improved toughness.

6.1.2 Thermoforming

The thermoforming industry can be divided into two distinctly different businesses. *Thin-gauge* thermoforming is devoted primarily to thin-walled, single-use packaging applications. *Heavy-gauge* refers to all kinds of industrial or nonpackaging applications. Rotational molding vies with heavy-gauge thermoforming for many of the same types of applications.

Thermoforming is a low-pressure, high-temperature, open-molding thermo-plastic process that converts heated, flat, two-dimensional sheets into larger, three-dimensional shapes with positive or negative air pressure.

Over the years, this industry has developed many different processes and special machines to fill marketplace opportunities. All but one of these processes is limited to producing open parts.

Twin-sheet and clamshell thermoforming are two similar processes that are capable of producing one-piece hollow parts. The growth of the twin-sheet thermoforming process that now dominates this part of the industry was hindered by restrictive patents issued in 1964. The expansion of this segment of the industry has been rapid following the expiration of those patents. This is a good indication of the increased interest in hollow plastic parts.

In its simplest form, twin-sheet thermoforming uses a specially designed machine equipped with a two-piece mold, or die, that defines the shape of the required part. Two sheets of thermoplastic material, cut to the required size, are placed in one or two clamping frames (Fig. 6.8). The machine sequences the clamping frames and plastic sheets through the heating chamber or oven. When the sheets have reached forming temperature, the clamping frames move the sheets to the forming station. The two halves of the mold close and press the two sheets together around the periphery of the cavities. The pressure of the molds welds the hot sheets together. Pressurized air is then introduced through the blow pin and the sheets are stretched to conform to the shape of the cavity. Vacuum may be used to form, or to assist in forming, the sheets. One or both of pressure and vacuum hold the hot sheet in contact with the cavity while the part cools. Cooling is achieved with water lines drilled through the body of the mold. After the part has cooled and regained strength enough to retain its shape, the molds

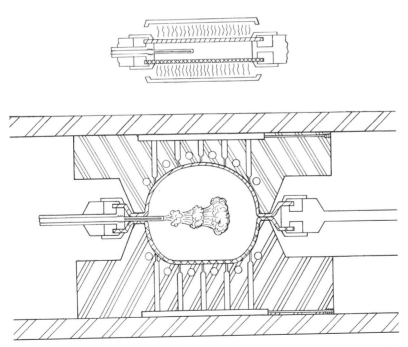

Figure 6.8 A schematic of the twin-sheet thermoforming process illustrating the heating of two sheets of plastic material and the forming of a hollow dock float

open. The machine then moves the clamping frames to the open station, where the clamps open and the formed part is removed. Two new sheets of plastic material are loaded into the clamping frames to initiate the next cycle. The excess sheet material between the clamped edges and the formed part is then trimmed away.

Twin-sheet thermoforming can be performed on a single-station machine where the basic steps of loading and clamping the sheets, and heating, forming, cooling, and unloading of the finished part are done sequentially. These machines are low in cost, but the production rate is slow. Single-station machines can be automated, but they are usually manually operated.

If the volume of parts to be produced justifies it, forming can be done on fully automatic multistation rotary machines. The sheet stock is loaded into the clamping frames by a robot. A second robot removes the finished part and places it on a conveyor.

A single-station machine can produce a 6.4 mm (0.250 in.) thick part on a ten- to fifteen-minute cycle. A highly refined, four-station rotary thermoforming machine is capable of producing one 5.6 mm (0.220 in.) thick part every 2.7 minutes. The rate-controlling part of the cycle is the time required to cool the formed part, which is in contact with the cooling surface of the die on only one side.

The multistation rotary thermoforming machines have many things in common with multi-arm carousel rotational molding machines. One advantage of thermoforming is that a multistation machine only requires one mold. To achieve the same production rate, a rotational molder would have to mount one mold on each individual arm of a multi-arm machine.

The largest commercially available twin-sheet thermoforming machines have a forming area of $4.6 \times 3.0 \times 1.2$ m ($15 \times 10 \times 4$ ft.).

For a given size, single-station thermoforming machines cost about the same as single-arm rotational molding machines. Multistation thermoforming machines cost approximately the same or more than multi-arm rotational molding machines.

Most thermoforming machines are run on electricity, but some heating chambers are now heated with natural gas.

Most production molds used for twin-sheet thermoforming are cast or machined aluminum. Their cost is approximately the same as a comparable size and shape of a two-part rotational molding mold. The cost of thermoforming molds increases sharply with an increase in size and complexity.

Both twin-sheet thermoforming and blow molding have benefited by being a part of the large packaging industry. The material manufacturers have pursued this large-volume business and have developed a wide range of materials that are suitable for these two processes.

Most common thermoplastic materials have been specially compounded to suit the thermoforming process. PE, ABS, PPO, PC, rigid PVC, polystyrene, acrylic, polysulfone, and the cellulosics are all routinely thermoformed for industrial applications. This wide selection of available materials is a distinct advantage of thermoforming over blow and rotational molding.

During the thermoforming process, the sheet of thermoplastic material is brought up to temperature in a heating chamber. The sheets are heated to a lower temperature than that used for rotational molding, but both processes subject the plastic material to the possibility of oxidation and thermal degradation.

The thermoforming process requires the plastic material to be in the form of a flat sheet. The majority of the work being done today is with extruded sheet stock. Large-volume thermoformers extrude their own sheet in-house. Small processors purchase their sheet from a company that specializes in sheet extrusion. Extruding, cutting the sheet to size, and the extra shipping can add 55¢ to 66¢ per kilogram (25¢ to 30¢ per pound) to the cost of a pelletized material.

Thermoforming is limited to working with the standard size, color, and types of materials that are available in sheet form. Sheet stock can be purchased in standard thicknesses of 0.38 mm (0.015 in.) to 12.7 mm (0.500 in.). Some 19.0 mm (0.750 in.) sheet is commercially available, but any thickness can be secured on a special order basis. There is no limit to the length of sheets that can be produced. The width is currently limited to 3.0 m (10.0 ft.).

Laminated and coextruded sheet stock is now available. Some successful work has been done with foamed sheets and laminates containing foam. One unique capability of twin-sheet thermoforming is its ability to produce a two-colored part by using a laminate of two colors of sheet stock. The bottom of a food industry pallet could be black and the top white.

Depending on the size and shape of the part produced, the material trimmed away from a thermoformed part is, on average, 30% but can be as much as 40% by weight. This material can be recycled, but there is an added cost associated with regrinding and reextruding the trimmings into new sheet stock.

Blow molded, rotationally molded, and thermoformed parts are all machined after molding to provide openings and other special features that cannot be molded in. The excess material cut out of these parts must also be recycled and included in a part's total cost. The material portion of a product's cost is normally the highest for thermoforming and the lowest for blow molding, with rotational molding somewhere in between.

Both thermoforming and blow molding are stretching processes. Depending on the shape of the part and the plastic material being used, there is a limit to how far the material can be stretched. Rotational molding does not have this limitation. Neither process can produce parts with the wall thickness uniformity that can be expected with rotational molding. The ideal shape for a twin-sheet thermoformed part is a ball. A ball-shaped part will exhibit the minimum variation in wall thickness. A six-sided square tank will have noticeable thinning at the corners.

Custom pallets, especially of the conforming and nesting variety, are being produced by both twin-sheet thermoforming and rotational molding. The initial investment in machines and molds is approximately the same for both processes. Large quantities of relatively shallow-depth, simple shapes favor thermoforming. Smaller quantities of complex-shaped, heavy-duty pallets tend to be rotationally molded.

The blow molding, rotational molding, and thermoforming processes are all capable of running multiple-cavity molds of a given part. All three processes can simultaneously run molds of different parts. Rotational molding is, however, the best process for simultaneously producing a mix of different parts.

Blow molding and twin-sheet thermoforming have the advantage of not requiring the use of a mold release agent.

During the rotational molding and thermoforming processes, the cavity is the only thing that touches the hot plastic. Both processes lend themselves to quick mold changes and setups. Blow molding, which melts the plastic material in an extruder, takes many times longer to set up a new mold and to change from one type or color of material to another.

6.2   Selecting a Hollow Part Process

Once a design engineer determines that a hollow part is advantageous, the next decision becomes which of the three hollow thermoplastic processes should be used. Each process has its own advantages and limitations, and there is a lot of overlap in the choice of a process. Commercially successful refuse containers have been produced by rotational molding, thermoforming, and blow molding.

If all three processes are capable of producing a functionally acceptable product, the next question becomes one of time and money. Mold costs and delivery for rotational molding and thermoforming are generally less than for blow molding. Blow molding normally has the lowest part cost, but the size, shape, and volume of product to be produced could favor any one of the three processes.

The type of plastic material to be molded is another consideration. All three processes work well with PE, but PP is a difficult material to thermoform. Rotational molding is the only one of the three that can mold the liquid PVC plastisols.

It is sometimes possible to arrive at the optimum process by elimination. The design engineer should always watch for some required product feature, such as size, shape, cost, or material, that will eliminate a process from further consideration. For example, the size of the part may exceed the capacity of current blow molding machines. The width of commercially available sheet stock is a limitation to the size of thermoformed parts. Rotational molding is capable of producing the largest hollow, one-piece parts.

The shape of a part, such as the sludge-lift tank (Fig. 1.4) cannot be produced as an integral part by blow molding or thermoforming. The lining of metal pipes, tanks, and valves is another example of parts that can only be made by rotational molding.

If the product includes closely spaced parallel walls, rotational molding is the first choice. Blow molding is also used to make parallel-walled parts, but the process requires a more complex and costly multiple-action mold.

Rotational molding is the best of the three hollow-part processes for molding in heavy, load-bearing inserts. It is also the best process for incorporating large inserts.

In those cases where a uniform wall thickness or strong outside corners are important, rotational molding is the best choice.

Products requiring a high degree of toughness have the best chance of success with rotational molding. This process's relatively low levels of molded-in residual stress improve a plastic material's impact strength and chemical resistance.

All three hollow part processes are capable of producing parts with multimaterial walls. For example, a thin layer of expensive chemically resistant material can be surrounded and strengthened by a thicker layer of a lower cost material. A foamed wall, in combination with one or two nonfoamed walls, is also possible. Rotational molding conveniently provides these capabilities with the minimum investment in equipment.

In the final analysis, the only justification for choosing one process over another is that it is the best manufacturing procedure for a project.

Some of the important attributes of blow molding, rotational molding, and thermoforming are summarized in Table 6.1. This table is generic in nature. It does not represent any one product. The ratings are the author's best estimates for a hypothetical part that could be molded by all three processes. The values assigned to each process and each attribute might change, depending on the size

Table 6.1   Attributes of Hollow Plastic Parts Processes

	Rotational molding	Extrusion blow molding	Twin-sheet thermoforming
Shape complexity	10	8	7
Maximum size	10	5	8
Minimum size	10	9	8
Thickness uniformity	10	6	7
Precision dimensions	6	10	5
Postmold warpage	6	10	7
Finished both sides	0	0	8
Finishing required	10	9	0
Undercut capability	10	8	7
Molded threads	9	10	7
Molded-in inserts	10	4	5
Inline closed holes	10	8	1
Inline open holes	10	1	7
Side-cored closed holes	10	6	1
Side-cored open holes	10	0	5
Material choices	6	8	0
Multimaterial capability	10	9	10
Foamed materials	10	1	8
Material scrap factor	10	8	7
Material cost	8	10	6
Machine cost	10	7	6
Mold cost	10	7	9
Labor input	6	10	9
Production rate	7	10	7
			8

10 = Best or most common   5 = Worth considering   0 = Normally not done

and shape of the part or the quantity being produced. In spite of its limitations, the table is useful as a quick reminder of the attributes of the three hollow part producing processes. The use of this table and the product development checklist (Table 2.9) are an inexperienced designer's best guide to selecting the ideal hollow part process.

The numerical values assigned each process-attribute combination are indicated as ten for the best or most frequently specified combination. A rating of five or above is worth considering. A value of one indicates that this combination is rarely specified, or that it can only be done with significant difficulty. Zero indicates something that is not normally attempted with that process.

7 Predictions for the Future

The rotational molding industry and in fact the whole commercial world is in a state of transition. Attempting to predict the future for this industry and its customers is a fool's game. No one actually knows what the future will be beyond the fact that it will be different. Most predictions are based on what happened in the past with an allowance for foreseeable changes. The rotational molding industry is now changing so rapidly that what happened in the past is not necessarily a reliable indication of what will happen in the future. Past performance is, however, more reliable than counting tea leaves. Recent and current developments in this industry do provide some insight into at least the immediate future.

7.1 The Industry

The rotational molding process and the business itself are unique. There is nothing quite like them in the rest of the plastics industry. In spite of these differences, rotational molding shares many of the same attributes as the rest of the industry.

Consumers purchase two-thirds of all products. The population of the world is increasing and it is becoming more affluent. In North America the growth in the use of plastic materials for the last forty years has been greater than the increases in gross domestic product. The growth of the plastics industry will continue through the foreseeable future, as plastic materials continue to capture the markets formerly enjoyed by other material-based industries.

As the number of worldwide consumers increases, the plastics industry will continue to grow. The use of plastic materials will increase at a faster rate than the rest of the economy. Rotational molding will grow right along with the rest of the plastics industry, but at a faster rate than most other plastics processing based industries.

All through the 1980s and the first half of the 1990s the rate of growth of the rotational molding industry was in the range of 10 to 15% per year. [10]. It is unreasonable to expect this kind of growth to continue indefinitely. A more realistic prediction would be for sustained growth of two to three times the gross domestic product. Average growth of this amount would be faster than the rest of the economy and the rest of the plastics industry.

The users of rotationally molded parts can expect to see more consolidation in the industry as larger molders purchase smaller molders. As these companies become larger, they will become full-service suppliers with molding plants in strategic locations across the U.S. These full-service molders will have the financial capability to adopt the very latest technologies. OEMs and their suppliers will benefit from this consolidation of the industry.

The success of the rotational molding business and the publicity spawned by its impressive growth rate have now attracted the attention of financiers. Investors are starting up new molding companies and purchasing existing processors. In most cases these start-up and acquired companies are managed by organizations that are new to the industry. Many of these chief executive officers do not have prior experience with the rotational molding process. For understandable reasons these people manage their companies from a financial or return-on-investment perspective. Their motivations are very different from the attitudes of the entrepreneurs who founded and for the time being still dominate this industry. This trend will continue for as long as the rotational molding business continues to generate better than average profits. This trend is changing the personality of the industry as it has with other parts of the plastics industry. As a result, the industry will become more professional, but some of the fun of doing business will be lost.

There will be an increase in the current trend for full-service rotational molders to offer other processes, such as blow molding and twin-sheet thermoforming. These suppliers will no longer be rotational molders. They will become hollow plastic part producers. Other processors will branch out into injection molding, thermoset reinforced plastic fabrication, and structural foam molding to become plastic part producers and not just hollow part molders.

In today's global economy OEMs are becoming accustomed to purchasing whatever they need from anywhere in the world where the cost, quality, and delivery are the most favorable. With the exception of relatively small parts that are economical to ship, the users of rotationally molded parts will continue to purchase these parts in the same country or same part of the country where they will be used. In other words, large rotationally molded parts will continue to be shielded from offshore competition.

Niche marketing has resulted in many formerly large-volume applications being converted into several smaller markets. At the same time plastic parts are becoming more complex and larger in size. Producing smaller quantities of large parts is ideal for a low-pressure process such as rotational molding, which requires a low initial investment in molds and molding machines.

It can be anticipated that large OEMs will continue the current trend of outsourcing more and more of their manufacturing requirements. This trend will change the structure of the rotational molding industry. According to a 1996 survey, 64.0% of the rotational molding business was devoted to producing proprietary products; 32.4% of the business was custom molding; captive and miscellaneous molding accounted for only 2.1% and 1.5% respectively [22]. This data reveals one of the unique aspects of this industry. Rotational molding companies are capable of developing, producing, and marketing their own products. This is different from the rest of the plastics processing industry, which depends on custom molding for the majority of its business. The changing purchasing practices of OEMs will result in an increase in the amount of custom molding that will be done in the future.

The rotational molding industry is in a state of transition. The more progressive molders have embraced the new technologies, while others have adhered to the status quo. Increased customer expectation and demand for continuous improvement in cost and quality will either upgrade or eliminate the less progressive molders. Either way, the industry and its customers will benefit in the years to come.

The biggest challenges for the immediate future are that the molders have to accelerate their acceptance of technological developments in order to improve the quality and reduce the cost of rotationally molded parts. The challenge for OEMs is to learn when and how to use this process to its maximum capabilities.

All things considered, the future of the rotational molding industry will be good for both OEMs and their molder suppliers.

7.2   Emerging Technologies

In the past, the bulk of the research directed toward rotational molding was conducted by the plastic material manufacturers. The research and development of improved PEs is continuing. With this one exception, most large material suppliers have now reduced their research activities in order to cut costs and increase short-term profits. Fortunately, this lack of support of the industry has been offset by the work being done at universities and in the private sector. Today there is more research and development work being performed than at any time in the past. An impressive array of technical innovations has already been made. Advances are now being made faster than the industry can, or is willing to, assimilate them. Considering the investigations now underway, it is reasonable to assume that technical breakthroughs will continue to be made. There are so many developments in all areas of the industry that they have to be categorized just to keep track of them all.

7.2.1 Advances in Plastic Materials

Rotational molding is a plastic material based industry. The applications that are available to the industry are dictated by the commercial availability of plastic materials that are suitable for the process. Some OEMs are prevented from specifying rotational molding because it cannot mold the material that they require. This limitation has been recognized and steps are being taken to overcome it.

The manufacturers of PE have been successful in providing the industry with continuously improved materials. The relatively new metallocene catalyst technology has already resulted in improvements in the uniformity and physical properties of PE. There is every reason to believe that additional advances will be made in the future.

The PPs that were reintroduced in 1995 are now increasing their market share. This increase in use will encourage other material manufacturers to enter the market with improved grades. The current suppliers will respond with even better grades allowing processors and their customers to capture additional markets. Following the trend established in other plastics processing industries, the use of PP will increase in the future.

ARM is continuing to pursue university research in the development of a new abrasion-resistant material. The search is also underway for a moldable grade of ABS and a PPO.

Shell Chemical Co. has introduced a polyketone material under the trade name of Carilon. This material has been successfully rotationally molded on an experimental basis. Shell is now concentrating on developing Carilon for the large-volume injection molding field. To date the applications have been similar to those now being produced in acetal and nylon. This material may develop into a new rotationally moldable material.

Both thermoplastic and thermosetting polyurethanes have been rotationally molded in the past, but large-volume applications have not developed. That situation is now changing. The huge automotive industry is beginning to show a preference for polyurethane head- and armrest skins. Impact- and abrasion-resistant toys and specialty tanks for the agricultural market are other growing polyurethane markets. If the transportation industry continues this trend, the size of the market will increase, which will encourage additional developments in polyurethane materials and processing techniques. Once these materials and processing techniques have been perfected, OEMs can specify them for other markets.

The cross-linkable liquid nylons are now finding their way into the rotational molding industry. Cross-linked Nyrim has been successfully molded as an attractive high-performance gasoline tank for a BMW motorcycle (Fig. 7.1). Additional applications are now under development.

The liquid thermosetting phenolics, epoxys, and especially polyesters have been successfully rotationally molded in the past. All of these cross-linkable materials can be heavily filled and fiber-reinforced. The epoxys and polyesters can be compounded to cross-link without heating. Room temperature curing

Figure 7.1 Rotationally molded, cross-linked nylon motorcycle fuel tank (Courtesy DSM RIM Nylon, Inc., Augusta, GA)

results in long molding cycles; however, these slower production rates have not stopped the growth of other thermosetting processes such as filament winding, lay-up, spray-up, and resin transfer molding. This capability allows the use of simplified molding machines that do not require heating and cooling chambers. The room temperature curing polyesters and epoxys have been rotationally molded in glass fiber reinforced, cast silicone rubber molds. This fast, low cost mold building technique allows a molder to build molds in-house and be in production in days instead of weeks. The inherent advantages of rotationally molding these thermosetting materials holds great promise for the future. In fact, it is hard to understand why these commercially available materials have not already found wider acceptance within the industry.

PE is an ideal material for the rotational molding process. Its combination of cost, properties, and ease of processing accounts for its 85% share of the market. If PE is to remain as this industry's material of choice, then major increases in stiffness and temperature resistance will have to be made. There is a limit to what can be achieved with the various members of the PE family. OEMs need a material that can be used in heavy load bearing applications. Future research needs to be directed toward finding ways to improve PE's stiffness and temperature resistance through the use of fillers and fiber reinforcements. The use of fibers and fillers has extended the use of other plastic materials molded by other processes. The rotational molding process should be able to do the same thing.

More and more materials will become available as micropellets. In the future more molding will be done with micropellets and blends of powder and micropellets.

The increasing use of single-shot foamed PE also holds great promise for the future. Foaming a PE produces a higher stiffness per kilogram (pound) of material. Many future uses will be found for this technology as the material developments continue and as molders gain additional processing experience.

A recent change in the overall plastics industry has resulted in a situation where major plastic material manufacturers serve their small customers through distributors. The emergence of these distributor networks has, in turn, led to the development of an increased number of specialty compounders. It was never, or rarely ever, possible to get a large material manufacturer to produce a small quantity of a specially compounded plastic material. The distributors and specialty compounding companies have now built a business based on filling these smaller volume needs. OEMs and their custom molder suppliers must not overlook this opportunity to recompound the commonly molded materials to meet special marketplace needs. The use of specially compounded materials will increase in the future.

The immediate challenge for the plastic material part of this industry is to develop additional materials that will allow OEMs to specify rotational molding for more demanding applications.

7.2.2 Advances in Product Design

The plastic product design community is, and will continue to be, under extreme pressure to create high-quality plastic parts that can be reliably produced at a low cost in the shortest possible time. One common way of reducing cost is to minimize molding and assembly operations by combining several components into one part. The metal-to-plastic conversion of the fuel tank cited earlier is a classic example (Figs. 6.1 and 6.2).

Design engineers are slowly becoming more familiar with the capabilities of the rotational molding process. As they gain additional understanding of the process, they are designing progressively more complex parts. In some cases these more complex parts are pushing the process to its maximum capability, but that is one way that progress is made. Engineers still design rotationally molded tanks, but the trend today and in the future will be to combine these tanks with other structural elements in the product. The one-piece farm tractor part (Fig. 7.2) connects the two floor-level fuel tanks with two hollow vertical side walls that provide ducting for ventilation and air conditioning, while providing the structural support for the instrument panel. Many other details, such as the tank-filling spout, are also molded in. Farm tractors are a demanding application, but good design allowed this part to be produced in LLDPE.

Another example of the benefits of using a hollow part process to reduce assembly costs by combining parts is the power sweeper frame (Fig. 7.3). This one-piece hollow frame replaced over 120 metal parts used in the previous metal sweeper. Molded-in color eliminates painting. A major advantage cited by the OEM is the elimination of the need to assemble over 100 parts after the frame is received. The sweeper frame, molded by Pawnee Rotational Molding Company, was awarded ARM's [13] 1997 State of the Art and Product of the Year awards. That same part won SPE's [9] 1998 Plastics Industrial Product Design award.

Both the farm tractor and the power sweeper frames are heavily loaded, complex parts. An understanding of the rotational molding process and meticulous attention to design details are required in order to produce commercially successful parts of this type in PE.

More and more design engineers are learning how to design properly proportioned plastic parts for the rotational molding process. At the same time

Figure 7.2 A PE farm tractor part that combines two fuel tanks, the tank-filler spout, floorboards, hollow vertical channels that provide heating and cooling air passages, and support for the instrument panel (Courtesy Rhein-Bonar Kunststoff-Technik GmbH, Germany)

many designers are using this process for the first time. There is nothing in the usual university engineering curriculum that teaches designers how to design parts for the rotational molding process. Fortunately, design guidelines are available (Chapter 3) to help the novice design engineers develop their first rotationally molded plastic product. Published design guidelines are helpful, but they cannot cover all situations or answer all a designer's questions. Consulting with an experienced molder during the development of a first-ever rotational molding project will enhance that product's chances of success. A recent survey indicated that 86% of custom rotational molding companies provide part design as one of their secondary services [22]. The current trend of molders assisting their customers with part design will increase in the future.

There is a trend for more rotationally molded parts to be designed using CAD. The three-dimensional solid models produced by CAD are highly desirable, because the design engineer is then in a position to use that same database with other emerging technologies such as the CNC machining of molds and finite element analysis (FEA). The increased use of FEA as a product design

Figure 7.3 A multifunctional, load-bearing PE sweeper frame. The fully assembled sweeper uses ten rotationally molded parts (Courtesy Tennant Co., Minneapolis, MN)

tool will allow OEMs to accurately develop heavy load bearing products while avoiding the common problem of overdesigning. FEA will also accelerate the rate at which other materials are replaced by plastics.

Computer-aided mold filling analysis software programs of the type commonly used for the melt flow processes have not been developed for rotational molding. Some research has been directed toward developing simulation programs for rotational molding, but none of these efforts has produced a commercially available program. This work is continuing and software programs of this type will become available in the future. When these programs are perfected, the designer will be able to use them in conjunction with CAD and FEA to model the entire design, development, and processing aspects of a project.

7.2.3   Advances in Mold Making

In the past ten years the mold making part of the rotational molding industry has contributed a multitude of small but important improvements in accuracy, quality, delivery, and tooling features. This part of the industry is now on the verge of landmark advances in mold technology. The increased use of CAD-generated three-dimensional solid model product design drawings or databases will continue. The use of computer-generated cutter path programs and the CNC machining of patterns, cavities, and whole molds will increase. In time, FEA programs will be capable of accurately predicting different mold shrinkage factors for different locations on a part. In the future, molds will be built allowing for large shrinkage factors on those surfaces that can pull away from the cavity during cooling. Smaller shrinkage factors will be used on confined details such as fixed inserts and surfaces that shrink onto cores in the mold. When all these computer programs are integrated, the industry will be able to significantly shorten its mold delivery schedules. This will be due to more efficient mold building and cavity finishing. The ability to properly size the cavity the first time will also allow new products to be introduced sooner.

The increased efficiency and accuracy provided by the CNC machining of patterns will encourage greater use of electroformed cavities for products other than figurines and doll parts. One of the four large injection molding tools being used for Chrysler's all plastic bodied Composite Concept Vehicle (CCV) is electroformed. It is intriguing to consider the advantages of rotationally molding that whole car body as one piece in an electroformed mold. Such a part could also incorporate the fuel and radiator overflow tanks. Ventilation ducts, instrument panels, and storage compartments are other possibilities. This would be an impressive parts consolidation project. The plastic material now being considered for the injection molded CCV is a thermoplastic polyester. What material could be specified for a rotationally molded CCV?

The sewn fabric industry is finalizing the development of software programs that will be useful in the engineering of fabricated molds. These programs are intended to improve the ability to cut two-dimensional sheets of fabric to conform to three-dimensional shapes, such as garments or upholstered furniture. These programs, or a modification of them, will allow fabricated molds to be economically built to more complex shapes.

The old concept of pumping hot and cold fluid through passages in a mold to quickly heat and cool the cavity and the plastic material is now being reconsidered. The Kelch Corp. debuted a direct heating and cooling mold of this type at a 1997 trade exposition. The improvements made over the years in the equipment used for heating, cooling, and conveying liquids is now being brought to bear on the rotational molding process. To cite just one example, the leakage and safety problems associated with pumping pressurized hot oil through rotary unions can be eliminated by pulling the oil through the system with a negative pressure pump. Some of the advantages of this technique are that only the cavity and none of the rest of the associated hardware is heated and cooled. Liquid is better than air as a heat transfer medium. These direct heated and cooled molds should be very energy efficient. Work in this area holds great promise for the future of certain kinds of parts, and especially for shapes that lend themselves to production on single-arm or rock-and-roll machines. It is quite possible that the improvements being made in internally heated and cooled molds will spawn the development of a new type of molding machine that does not require heating and cooling chambers.

Another breakthrough in mold construction is the composite mold technology (CMT) developed by Andrew Wytkin in Australia (Fig. 7.4). CMT molds are built of glass fiber reinforced epoxy. Electrical heating elements and cooling channels are embedded in the composite cavity wall. Controlling the electric heating in different zones of the mold is possible. This differential heating would allow the molding of a cylindrical tank with thin walls at the top and a thicker wall near the bottom where the loads are higher. The bottom inside corners of deeply recessed double-walled hollow parts could be preheated to eliminate the thinning of the walls in those hard-to-heat locations. Another intriguing possibility would be to preheat a cavity in a narrow band that would produce a narrow but thicker reinforcing rib on the inside of the part while maintaining a smooth uninterrupted surface on the appearance side of the part. This would be something that the conventional heating of molds has not been able to accomplish. CMT claims to be capable of producing molds at a reduced cost and delivery. CMT molds are brand new and they have not yet had time enough to prove their production worthiness. It is the opinion of the author that CMT molds will become a standard part of the rotational molding industry in the future.

The advantages of molding with a positive gas pressure have now been accepted. The use of an inert gas to purge a cavity of oxygen and to force a moulded part to remain in contact with the cavity during cooling is increasing. Research work at Queen's University in Belfast, Northern Ireland, indicated that a positive gas pressure can be used during the molding cycle to accelerate the elimination of bubbles. That same work revealed that flowing air through a molded part during the cooling cycle can reduce molded-in stress by forcing the inside surface of the part to cool at a rate nearer the rate of cooling on the outside surface of the part. Internal cooling and the rapid removal of bubbles are two new molding techniques that are not yet fully proven, but they do hold the promise of significantly reducing cycle time and part cost.

In the future, molding with a positive gas pressure will become common. The increased use of a positive gas pressure will affect the way molds are designed and built. Molds can be made strong enough to withstand higher pressures by simply making the walls of the cavity thicker. Thicker walled cavities are, however, undesirable because they increase the weight of the mold.

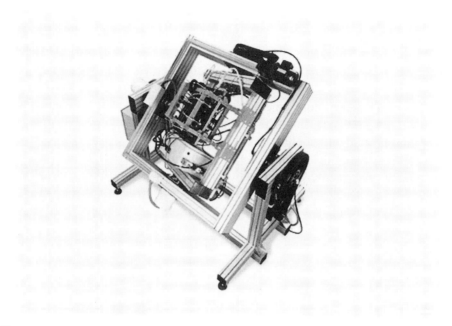

Figure 7.4 The Wytkin CMT direct heat composite mold mounted in its all-electric rotational molding machine that eliminates the use of heating and cooling chambers (Courtesy Chroma Corp., McHenry, IL)

Thicker cavity walls also take longer to heat and cool. Mold makers are going to be under increasing pressure to find ways of building stronger molds that do not increase their weight or cycling time.

Automating the opening and closing and loading and unloading of rotational molding molds has lagged behind the rest of the plastics processing industry. Techniques have been developed for automating all the thermoplastic molding procedures. A few large-volume rotational molding projects in industries such as automobiles and toys have been fully mechanized (Fig. 5.11). There are many complicated reasons why the rotational molding process has not been automated, but these obstacles will have to be overcome for this industry to survive in the future. The reduction or elimination of labor costs and the cycling variables associated with manual operation have been proven to be beneficial with other competitive molding processes. The same benefits are achievable with rotational molding.

The mold-making industry has been diligently working at reducing the time and labor required to open and close molds. Fast-acting clamping mechanisms are now available; others are under development, but they are all manually activated. The level of precision provided by CNC-machined molds holds the promise of simplifying mold clamping. The internally heated and cooled molds that do not subject a mold to a hot oven or a wet cooling chamber simplify the provision of fully automated mold closing, clamping, unclamping, and opening. All indications are that the industry is approaching a breakthrough in mechanizing this part of the mold cycle.

Leader pins and bushings are widely used in other molding processes to properly align the various parts of a mold during the mold closing process. These alignment devices will find wider use in rotational molding in the future as a method of eliminating parting line–flange damage and downtime for mold maintenance.

Automated liquid- and powder-conveying and metering equipment has already been perfected. The use of this equipment to mechanize the charging, or filling, of molds is simply a matter of cost benefit and personal preference.

Mold makers have been creative in developing devices to simplify the opening of molds and the removal of parts. Demolding has always been problematic. Some parts are large, heavy, and awkward to handle. It is a common practice to mold more than one size and shape of part on the same arm. Improper application and maintenance of mold release can result in a part sticking in the cavity. Many parts are designed without draft angles; others contain undercuts that complicate demolding. A few molds are designed to incorporate part ejection mechanisms. The added labor, increased cost, cycle-time interruption, and potential for mold damage will, in the future, mandate greater use of pneumatic poppit valves and all kinds of mechanical ejection mechanisms of the type now being used with other plastics processes.

The rotational molding industry and its customers will all benefit from the improvements being made by mold makers. The final automation of the whole process cannot be achieved by mold makers alone. Machinery manufacturers have to provide molding machines with the capability of automation, and molders have to purchase this new equipment. Customers must continue to encourage their suppliers to become more efficient by demonstrating a willingness to share in the increased cost of setting up an automated system that will generate greater profits over the life of the project.

7.2.4   Advances in Processing Machines

Every segment of the rotational molding industry is changing for the better. Product designers are creating increasingly larger and more complex parts that are pushing the state of the art. At the same time, customers are demanding and getting higher quality, lower cost, and shorter delivery schedules. These forces have encouraged the development of new and improved plastic materials. Excellent quality, easier to operate molds are now available for anyone astute enough to purchase them. Processing machines continue to become more efficient and easier to control, operate, and maintain. All these technical improvements are being brought together by the skill and knowledge of molders. The biggest changes in rotational molding both today and tomorrow will be realized on the molding room floor and in the parts being produced.

New styles of machines will continue to be introduced for the molding of specialty parts. For example, the special molding machines created to run the internally heated and cooled Andrew Wytkin CMT molds do not require heating and cooling chambers, and as a result are small in size (Fig. 7.4). Only the mold is heated and cooled. Eliminating the heating of the rest of the machine allows the automatic sequential charging of the mold with up to four different solid and foamable plastic materials at any time during the cycle.

Impending legislation resulting from the global warming crisis will result in increasing energy costs. This will, in turn, encourage a trend toward more energy-efficient rotational molding machines. The more compact and more efficient ovens associated with the single-arm machines will result in their increased usage. Internally heated molds of the hot oil and CMT types that do not require the heating of an oven, machine arm, spider, and mold frame will become more popular due to their lower energy requirements. The more efficient heating and cooling of molds, coupled with the preheating of the plastic material, will result in further reductions in cycle time.

The use of infrared heating will increase in the future. This method of heating molds can be very efficient for certain shapes of parts. In one case, an infrared-heated machine reduced a thirty-minute cycle to only 22.5 minutes. There is at least one infrared-heated rotational molding machine now running in the United States. There are reported to be five machines of this type now being used in France.

More and more rotational molding companies are becoming ISO certified. Statistical process quality controlled production is increasing in popularity. The extended capability of the newer computer-controlled molding machines encompasses an array of processing and quality data acquisition capabilities along with a bar code label printing feature. These labels can be attached to a molded part as a temporary or permanent record of the processing conditions that produced that part.

Automatic conveying, weighing, and dispensing of both powdered and liquid materials will be common in the future. An increasing number of the larger molders will reduce their material costs through in-house grinding. In-house pulverizing capability will reduce the cost and encourage the use of more postindustrial and postconsumer material. Customer demands for improved and uniform color consistency will result in the use of more high intensity mixers and wider usage of precompounded colored material.

Nearly all the molding machines being built today are equipped for molding with a positive gas pressure. In spite of this fact, the number of molders using gas is estimated at only 25 to 35%. The inherent advantages of using a positive gas pressure to eliminate bubbles and to shorten cooling cycles with internal cooling, while holding the molded part in contact with the cavity, are too good not to be adopted by all molders.

Cavities are now being purged with nitrogen to allow molding in an oxygen-free environment to prevent the oxidation of materials such as Nylon 6. Other nontraditional materials may also benefit from this technique. An underused capability of inert atmosphere molding is an improvement in the physical properties of PE. In some cases it will be possible to take advantage of these improved properties to reduce wall thickness and cost by using less plastic material and a shorter molding cycle.

In the future there will be increased use of the Rotolog, not only as a diagnostic tool, but also as a method of optimizing cycling conditions [23]. The advent of internally heated and cooled molds and the elimination of cooling and heating chambers will allow the Rotolog to communicate with the molding machine controls during production.

The infrared thermometry-sensing system previously discussed is now being evaluated under actual production molding conditions. The results to date are all positive. There is a high probability that this mold temperature sensing system can be correlated with cavity air and plastic material temperatures, while signaling the molding machine controls to adjust the molding conditions during the cycle. When this comes to pass, the rotational molding industry will have closed loop cycle control capability. This technology can be used to improve quality while reducing cost.

Blow molding, twin-sheet thermoforming, and rotational molding are competitive hollow part producing processes that are limited in their ability to produce molded-in open holes. All three processes require secondary machining operations to produce small holes, large openings, critical dimensions, and other features. Many of these features are provided by routing. The rotational molding industry has lagged behind the other two processes in adopting CNC three- and five-axis routers with or without tool-changing capability. Once a molded part is positioned in one of these units, all the subsequent machining operations are performed automatically. The use of CNC routers increases accuracy and reliability while reducing cost. There is no question that those molders who perform routing operations of this type will be forced to adopt these automated routers in the near future.

Technical advances are being made in all segments of the rotational molding process. These improvements will allow this process to continue to produce progressively more complex load bearing parts with higher quality and lower costs. All things considered, the future looks good for both the purchasers of rotationally molded parts and the industry itself. All indications are that this industry and its customers will continue to prosper throughout the foreseeable future.

References and Resources

1. Bregar, B., *Plastics News* (May 15 1995), Akron, OH, Phone (330) 836-9180
2. Copeland, S. B., *Rotation* (Fall 1996) Volume V, pp. 14–17
3. Paulson, D., *Plastics Design & Processing* (July 1970)
4. O'Connell, J., *Rotation* (Fall 1996) Volume V, pp. 23–29
5. Wilson, G. D., In *Development of Plastics*. Mossman, S. and Morris, P. (Eds.) (1994) The Royal Society of Chemistry, Cambridge, England, pp. 70–86
6. Bucher, J., In *Success Through Association.* (1996) Bucher Communications, Denver, CO., pp. 31–38
7. *Rotation* Magazine, JSJ Productions Inc., 4100 Westheimer, Suite 101, Houston, TX 77027, Phone (713) 621-3903
8. University of Wisconsin, Milwaukee, 161 W. Wisconsin Ave., Milwaukee, WI 53203, Phone (414) 227-3139
9. Society of Plastics Engineers, 14 Fairfield Dr., Brookfield, CT 06804, Phone (203) 775-8490
10. Mooney, P., In *Analysis of the North American Rotational Molding Business* (1997), 695 Burton Rd., Advance, NC 27006, Phone (910) 998-8004
11. Kliene, R. I., In *Rotational Moulding of Plastics*. Crawford, R. J. (Ed.) (1996) Research Studies Press Ltd., Somerset, England, pp. 4–7
12. Petruccelli, F., In *Rotational Moulding of Plastics.* Crawford, R. J. (Ed.) (1996) Research Studies Press Ltd., Somerset, England, pp. 69–71
13. Association of Rotational Molders, 2000 Spring Road, Suite 511, Oak Brook, IL 60523, Phone (630) 571-0611
14. D.A.T.A. Business Publishing, 15 Inverness Way East, Englewood, CO 80155, Phone (303) 799-0381
15. *Modern Plastics* Magazine, PO Box 602, Hightstown, NJ 08520, Phone (609) 426-7070
16. IDES, Inc., 209 Grand Ave., Suite 500, Laramie, WY 82070, Phone (307) 742-9227
17. Throne, J. L., In *Polymer Powder Technology*, Narkis, M. and Rosenzweig, N. (Eds.) (1995) John Wiley & Sons, Inc., New York, NY, p. 316
18. Beall, G. L., In *Rotational Moulding of Plastics*. Crawford, R. J. (Ed.) (1996) Research Studies Press Ltd., Somerset, England, p. 165
19. Scaccia, S. F., In *Rotational Moulding of Plastics*. Crawford, R. J. (Ed.) (1996) Research Studies Press Ltd., Somerset, England, p. 143
20. LaMont, B., In *Basic Principles of Rotational Molding*. Bruins, P. F. (Ed.) (1971) Gordon & Breach Scientific Publishers, New York, NY, p. 140
21. Taylor, T. J., In *Rotational Moulding of Plastics*. Crawford, R. J. (Ed.) (1996) Research Studies Press Ltd., Somerset, England, pp. 125–126

22. Grace, R., *Plastics News* (August 18, 1997) Akron, OH, Phone (330) 836-9180

23. ROTOLOG, Ferry Industries, Inc., 1687 Commerce Dr., Stow, OH 44224, Phone (330) 688-4400

24. Lycas, J., In *Rotational Molding of Plastic Powders.* (1975) Headquarters, U.S. Army Material Command, Alexandria, VA 22333, pp. 3–8

25. L & R Enterprises, 735 Sneed Rd. W, Franklin, TN 37069, Phone (615) 662-2377

Glossary of Common Rotational Molding Terms

The following list of terms is limited to words that are routinely used in the rotational molding industry. No attempt has been made to compile a complete list of terms used by the rest of the plastics industry.

Uniform terminology is important to good communication. This glossary is offered in good faith, but there are always exceptions. These definitions have been collected from many sources including the undocumented trade jargon. The author believes the definitions to be correct, but no guarantee is implied. Different segments of the industry from different parts of North America and the rest of the world use different words to define the same thing.

A **abrasion resistance** Ability of a material to withstand a mechanical action such as rubbing or scraping.

ABS Acrylonitrile butadiene styrene

absorber See UV (ultraviolet) absorber.

Acrylonitrile Butadiene Styrene (ABS) A terpolymer resin.

air release (plastisol) Rate at which air bubbles introduced during plastisol processing come to the surface and break.

alloy (mold or plastic) A blend of materials to obtain better physical properties than one material alone. Used commonly when referring to material used to make molds (e.g., aluminum alloy) or to a blend of polymers.

ambient conditions The conditions of temperature and humidity at a given location. Sometimes erroneously used to indicate standard conditions. See standard laboratory conditions.

American Society for Testing and Materials (ASTM) A body that standardizes testing procedures and sets specifications for materials.

amorphous Noncrystalline, without definite molecular order (e.g., general purpose styrene).

angle of repose (powders)   Included angle of the sides of a cone of powder formed as a certain amount of powder flows from a pourability funnel set at a certain height. It gives an indication of the flow characteristics of a ground powder.

annealing (mold) Heating, then gradually cooling to relax internal stresses in the material. This is commonly done in mold materials to achieve consistent dimensions after the heating and cooling process used in rotational molding. (plastic) A method to relieve stresses and strains that may be introduced during molding or during a postmolding operation.

antioxidant   An additive that inhibits oxidation degradation of the resin during processing, fabrication, and end use.

antistat   An additive used to eliminate or reduce static during the rotational molding process. Also used to reduce pigment swirling.

antistatic agent   Additive in resin or substance applied to the surface of a molded part to eliminate or reduce static charge.

arm   Mechanism that provides the mechanical support and the 90° biaxial rotation to the attached mold or spider.

ASTM   American Society for Testing and Materials.

automatic sieve shaker   An automated piece of test equipment that shakes and taps a series of different sized sieves in order to test a powder for performance size distribution.

axis   A straight line about which a mold rotates.

B **backbone**   Main chain of a polymer molecule.

baffle   A device to restrict or divert the passage of material. Often used to divert oven hot air into recesses in molds.

batch rotational molding machine   Type of machine where the oven and cooling chamber each has a fixed-position rotating arm. The molds are pushed manually from oven to cooling chamber, generally over roller conveyors, and attached to the arm by quick-lock connectors.

biaxial rotation   Rotation occurring simultaneously in two perpendicular planes.

bleed   Movement of additives, such as the plasticizer in a vinyl, to the surface of the molded part and often onto an adjacent part. See migration.

(color)   Movement of color to the outside surface of a polyethylene part.

(vinyl)   Movement of additives such as the plasticizer in a vinyl to the surface of the molded part and often onto an adjacent part.

blend   Mechanical mixture of two or more components.

blending resin   (PVC) A powdered vinyl polymer used to alter the flow or other properties of a plastisol, which is not suitable for use as the sole resin in plastisol.

blister   A raised area on the surface of a molded part caused by the pressure of gases inside it before the surface has cooled and hardened.

bloom   A whitish powdery deposit or a cloudy appearance on the surface of a part.

blow hole   A void through a molded part wall caused by pores or holes in welds, poor insert fit, porosity in aluminum casting, poor mold flange condition, or plugged or improperly sized vent.

blowing agent   Chemicals added to resin that on heating generate inert gases, causing the product to assume a cellular structure.

bluing in   A method of matching mold-parting lines to obtain a close fit.

boss (mold)   A projection from a mold surface, commonly used to produce an indentation in a molded part. (part) A protrusion, land, or bump on a part. A boss is used for such functions as locating, fastening, or adding strength.

bottom plate   See mounting plate.

branched   In molecular structure of polymers, refers to side chains attached to main (backbone) chain.

bridging   Improper flow of powdered resin in the mold due to resin touching across one wall to another. This condition can be caused by narrow mold passages, poor mold or part design, improper rotation, poor powder pourability, and improper resin type.

brittleness   A type of failure that occurs with low absorption of energy. Failure by cracking or shattering.

brittleness temperature   The temperature at which a resin changes from a flexible to a brittle state.

Brookfield   Instrument used to check viscosity of liquids. Readings are in centipoise. Water at one centipoise is the reference.

Brookfield viscosity   The viscosity of a liquid (plastisol) as determined with a Brookfield viscometer. The viscosity is reported in centipoise.

bubble   A spherical, internal void within a molded part. Differs from a blister in that it does not protrude at the surface.

bulk density   The weight of a material per unit volume. Gives an indication of the weight of powder that can be used in a given mold. Usually expressed in pounds per cubic foot.

C   **CAD/CAM**   Computer-aided design/computer-aided manufacturing.

carousel rotational molding machine   A machine that has two or more arms oriented in a horizontal plane around a central pivot point, allowing the three processing functions (heating, cooling, and loading/unloading) to occur simultaneously.

cast   To form a mold by pouring a liquid material into a cavity. The term is commonly used in referring to cast aluminum molds.

casting   The finished product of a casting operation.

catalyst   Additive that accelerates a chemical reaction without itself undergoing reaction.

cavity   The space inside a mold that forms the molded product.

centipoise   A unit of viscosity, conveniently and approximately defined as the viscosity of water at room temperature.

chalking   A powderlike residue on the surface of a material, often resulting from degradation.

charge   The amount of material by weight put in a mold to obtain the desired wall thickness of a finished part. See also shot weight.

chart recorder   (circular for oven profiles) A device that records temperature versus time of an oven.

chemical resistance   The resistance of a plastic to chemicals or compounds over a range of temperatures. Generally evaluated in terms of changes in dimension, weight loss, loss of physical properties, etc.

chemically foamed plastics   Plastic foam formed from a chemical reaction of its components, usually a chemical blowing agent.

clamshell rotational molding machine   A machine whose oven/cooling chamber doors function in a manner similar to a clamshell.

clamp   Device used to hold mold pieces together, but permit opening the mold for part removal during production.

clamping pressure   The pressure required during the molding operation to keep the mold closed.

closed-cell foam   The property of a foam of having each bubble completely sealed off from its neighbor so that no exchange of gas can take place except by diffusion through the walls.

CNC   Computer numerically controlled.

coefficient of thermal expansion   The linear expansion of a material with increasing temperature, expressed in units of cm/cm or in./in. per °C or °F. Values for common plastics range from 0.00001 to 0.0002 in./in. °C. This is measured using ASTM D-696.

coining   See pockmarks.

collapse   Contraction of the walls of a part usually upon cooling, leading to a permanent indentation.

color concentrate   A resin and high concentration of colorant, up to 50 or 60%, used to color natural resin. Usually it is in pellet form and is not normally used in the rotational molding process.

colorants   Pigments or dyes used to internally color plastics during the molding cycle. Colorants are available as powders, liquids, and pellets.

colorimeter   Instrument used to judge colors used at 1 to 5% levels by illuminating a sample with light from three primary-color filters.

composite   A material in which two or more distinct structurally complementary substances combine to produce some structural or functional properties not present in any individual component.

compound   An intimate admixture of a polymer with all materials necessary for the finished product.

compounding   The process of melt-mixing a polymer with all materials necessary for a finished plastic. Generally accomplished with an extruder.

compressive strength   Maximum crushing load at failure of a specimen divided by the original sectional area of the specimen.

Computer-aided design/computer-aided manufacturing (CAD/CAM)   The use of computer software to aid in the design of a part and then to control fabricating machinery in the manufacturing of the part or mold.

continuous rotational molding machine   A machine that has two or more arms, allowing the three process functions (heating, cooling, and loading/unloading) to occur on a continuous basis.

cooling   When all of the molding material has been properly fused, the mold is cooled by air or water spray while still being biaxially rotated.

cooling chamber   Generally an enclosed chamber that supplies cooling to the rotating mold through water spray or forced-air cooling.

cooling fixture   A device to hold a warm molded part in the desired dimensions and tolerances while it cools.

copolymer   A polymer formed from the reaction of two or more monomers.

core pin   A pin used to form a depression (on the outside of a rotationally molded part), which is an undercut to the way the mold is pulled apart. Core pins are removed before the basic mold is pulled apart.

coupling agent   A chemical additive that acts as the interface between the polymer and the glass fiber or other reinforcing filler to form a chemical bridge between the two. This improves the efficiency of the reinforcing agent, and higher physical properties may be achieved.

cradle (carriage)   A mechanism that supplies biaxial rotation to a mold or spider and supports it at two points.

crazing   Development of fine cracks on the surface of a plastic, sometimes extending into the body of the material.

creep   The permanent deformation resulting from prolonged application of stress below the elastic limit. Due to its viscoelastic nature, a plastic subjected to load for a period of time tends to deform more than it would from the same load released immediately after application, and the degree of this deformation is dependent on the load duration. Creep at room temperature is sometimes called cold flow.

crocking   A color deposit left when certain plastisol articles are rubbed against another surface.

cross-linked polyethylene   A polyethylene with chemical bonds between the molecular chains.

cross-linking   In polymer molecules, refers to the establishment of chemical links between the molecular chains. When extensive cross-linking occurs, the molecular weight increases, with the resulting improvement in physical properties.

cryogenic grinding   Reducing the particle size of plastics in the presence of a super-cold coolant such as liquid nitrogen or dry ice. This type of grinding is useful for very high melt or soft plastics.

crystalline   Polymer having regular order on the molecular scale. See semicrystalline.

crystallinity   A state of molecular structure in some resins that denotes uniformity and compactness of the molecular chains forming the polymer. Can normally be attributed to the formation of solid crystals with a definite geometric form.

cure   To change the properties of a plastic by chemical reaction; generally accomplished by the action of heat or catalyst.

cure temperature   The temperature that must be reached to activate the chemical reaction that causes a change to occur in the properties of a resin.

cycle   The cycle or cycle time of a molding operation is considered from one point to the corresponding point in the next repeated sequence.

D   damper   A baffle located in an oven at the burner entrance that is activated to reduce heat loss during oven door opening and to control the heat exchange rate to the rotating molds.

dart impact   Impact obtained by striking a part or a sample cut from a part with a falling dart.

deaerate   To remove the entrapped air from a vinyl plastisol by subjecting it to a vacuum.

decomposition products   The chemical by-products produced by decomposition of a resin or an additive therein such as a cross-link or foam initiator.

deflashing   Covers the complete range of operations used to remove the flash from a molded plastic part.

deflection temperature   The temperature at which a standard specimen of a material deflects 2.54 mm (0.010 in.) under a load of 0.455 or 1.82 MPa (66 or 264 psi) (ASTM D-648). Provides a rough estimate of the maximum end-use temperature of a plastic.

degradation   Deleterious change in a resin's chemical structure reflected in its appearance or physical properties; normally due to exposure to heat, light, etc.

degree of cross-linking   The number of chemical links between the molecular chains in a cross-linked polymer. When the degree of cross-linking increases, the cross-linked molecule becomes larger, giving the polymer a higher molecular weight. One indication of the degree of cross-linking for polyethylene is the percent gel. Although higher percent gel indicates a higher degree of cross-linking, the percent gel is not the actual percent of cross-linking that has occurred.

delamination   The splitting or separating into layers of a molded part wall. Can be caused by contamination by incompatible resins, rotation problems, or poor adhesion between multilayer products.

demolding   The act of removing parts from a mold during the molding process.

densification   Often referred to as sintering but may relate to the later part of the sintering process.

density   Ratio of the weight to the volume of material; many times expressed as pounds per cubic foot. In the SI system, the units are grams per cubic centimeter, which results in a density equal to the specific gravity. ASTM test methods are D-1505, D-792, or D-4883.

desiccant   A substance that can be used to attract moisture and is sometimes used during mold storage to prevent rusting.

diffusion   A slow movement of matter caused by a concentration gradient, such as the absorption of water or the transmission of gas by a plastic.

dimensional stability   A property of the material to hold its exact shape after it has been molded. Dimensional stability can be correlated to shrink factor and other material properties.

direct drive   A positive method for driving the mold-rotating mechanism.

discoloration   A change from the original color of a material; caused by such things as degrading the material by overheating, chemical attack, or exposure to sunlight.

dispersant   (PVC) A liquid component that has a solvating action on a resin, so as to aid in dispersing and suspending it.

dispersion   High shear distribution of one material in another (e.g., a vinyl resin sheared into a plasticizer during the compounding of plastisol, or dry color sheared onto powdered resin before molding).

dispersion resin   A fine particle homopolymer or copolymer vinyl resin that can be used as the sole resin in a plastisol because of its ability to form a dispersion with a plasticizer.

distribution   Uniformity with which pigments and additives are mixed with the polymer. Poor color distribution may appear as swirling or blotches of natural or uncolored resin.

double-cavity   A mold that makes two parts. Commonly used when a molded part with a large cutout area can be combined with a second part so that the cutout area is not formed.

double-wall   A part design that contains two close together walls, forming a single-walled structural component. This design may contain kiss-offs or be filled with foam for added stiffness.

draft   A taper designed into the mold. Commonly used to facilitate removal of a molded part from the mold. Draft is normally designated in degrees.

drill bushing   A hardened material through which a drill is normally inserted for location purposes. Drill bushings are commonly used to locate core pins, insert holders, etc., in molds to prevent wear on softer mold materials.

drill point   A small protrusion, or dedent, in the mold that leaves a small indentation on the piece part. It is used to locate the point of a drill when drilling holes in secondary operations.

drop box   An insulated box that is mounted to or inside a mold and holds a charge or charges of powder to be discharged inside the part during the heating cycle to produce a multilayered part wall.

drop-top or drop-out   A method of producing a thin wall or large opening in a molded part by insulating that portion of the mold from the heat of the oven.

dry blending   The mixing of ingredients by purely mechanical means without melting of the polymer.

dry color   Colorant and additives supplied in powder form. They are generally not resinated.

dry coloring   The process of combining plastic powder or pellets with dry color by mechanical action such as high-intensity mixing without melting the resin.

dry flow   A test used to predict the ability of a powdered resin to flow in a mold during rotation. A predetermined amount of resin is passed through a funnel with a specified finish and orifice. The faster it flows through the funnel, the better it should flow in the mold (ASTM method D-1895).

dryer   Equipment that removes moisture from hygroscopic materials before processing.

dyes   Transparent or translucent colors that are soluble in plastics.

E   **elastomer**   Molecules within plastics and rubbers that can be stretched several hundred percent and subsequently experience nearly complete recovery.

electroforming   A process to produce molds in which a relatively thin layer of metal is electrically deposited onto a pattern. The metal is sometimes backed up with other materials for added strength.

electroplating   A process to deposit a thin layer of metal over a surface.

elongation   The ability of the material to stretch without exceeding its break strength. The value is expressed in percentage of stretch within a given portion of the specimen (ASTM method D-638).

environmental stress-crack resistance (ESCR)   The ability of a resin to resist surface cracking under an induced load in an igepal (detergent) solution. The value is expressed in hours when 50% of the samples fail (ASTM D-1693).

environmental stress-cracking (ESC)   The susceptibility of a thermoplastic article to crack or craze under the influence of certain chemicals and stress.

ESC   Environmental stress-cracking.

ESCR   Environmental stress-crack resistance.

ethylene copolymers   Polymers formed by copolymerizing ethylene with polar monomers (e.g., vinyl acetate, ethyl acrylate, acrylic acid) or, as with the low-pressure resin, ethylene and alphaolefin monomers (e.g., propylene, butene, hexene, etc.).

exotherm   Extra heat generated by the chemical reaction of two or more components when mixed together (e.g., polyurethane or polyester thermosetting systems).

extender   A substance (usually a filler) added to a plastic composition to reduce cost. Other properties may also be developed.

extruder   A machine used to hot-melt compound plastic materials. The extruder melts the polymer through shear and heater bands on the barrel.

extrusion compounding   The dispersion of additives into a polymer using an extruder. The extruder melts the polymer and supplies shear to complete the dispersion.

F   fabricate   To produce a mold by machining, forming, and welding or fastening metal together. In general, fabricated rotational molds are made out of cold or hot rolled steel, stainless steel, or plates and sheets of aluminum. Fabricated molds are contrasted with cast aluminum molds.

FDA   United States Food and Drug Administration.

FD&C   Food, drug, and cosmetic.

fatigue   The failure of a material after repeated applications of stress, usually at high frequency.

fatigue life   The number of cycles of deformation required to bring about failure of a material under a given set of oscillating conditions.

feedstock   A base resin that is used in formulating a molding compound.

ferris-wheel rotational molding machine   A machine that has two or more arms oriented in a vertical plane around a central pivot point, allowing the three process functions (heating, cooling, and loading/ unloading) to occur simultaneously.

filler   A material added to a resin, usually in substantial percentages, to lower the cost or modify mechanical properties. Common fillers are calcium carbonate and talc.

fines   Very small particles (usually under 100 mesh) accompanying larger particles in a powdered resin.

finish   Commonly refers to the quality or texture of the surface of a mold that forms the part. The finishes commonly used in rotational molding are machined, ground, blasted (using sand, shot, etc.), etched, engraved, and plated.

fitting in   Matching parting lines to obtain a close and tight fit.

fixed-arm rotational molding machine   A rotational molding machine that has fixed-arm locations connected to a central hub; generally referred to as turret or carousel.

flame retardant   Reactive compounds and additive compounds incorporated into a resin to reduce its flammability.

flame-treating   A method of passing an oxidizing flame over a polyethylene part to change the surface polarity and allow the surface to be receptive to adhesion of inks, paints, etc.

flammability The relative burnability of the material in a specified situation. Meanings vary according to the test methods used (ASTM D-1692, ASTM E-84, ASTM E-162).

flange The metal added to a parting line area of a mold to strengthen the parting line and permit the use of bolts and clamps to hold the mold together during molding.

flash The extra plastic attached to a molding along the parting line of the mold.

flash line The line on a molded part that occurs at the parting line of the mold.

flexural modulus The ability of the resin to resist bending under load, and generally an indicator of the rigidity of the material. The value is expressed in MPa or psi (ASTM method D-790).

flexural strength The strength of a material in bending, expressed as the tensile stress of the outermost fibers of a bent sample at the instant of failure, or at 5% strain if the material does not break at the outer fiber.

flow rate The mass of material (in grams) that will flow from an extrusion plastometer of standardized dimensions under specified conditions of temperature and load in ten minutes. A flow rate run on polyethylene at 190 °C (374 °F) using a 2160 g (0.982 lb.) load is typically referred to as a melt index.

fluoroplastic A polymer containing fluorine, such as polytetra fluoroethylene or fluorinated ethylene-propylene copolymer.

fogging The release of incompatible or volatile ingredients from a plastic article. These ingredients then settle or condense on surrounding substrates.

follow board A board used to develop a parting line during the cast aluminum mold making process.

frame The support members around a single or multicavity mold that permit it to be mounted to a machine. The frame also strengthens the mold so it can withstand clamping and molding pressure, and permits the mounting of clamping and demolding devices.

fusible Capable of being melted or rendered fluid by heat.

fusion The processing stage where a plastisol forms a homogeneous solid during the application of heat. When the plastisol, after gelation, is fused, full physical properties are developed.

fusion temperature The temperature at which fusion occurs.

G **Gardner Impact** Instrument used to measure impact strength of a plastic using a falling weight from a predetermined height.

gas injection The use of inert gas or air injected into a mold to prevent oxidation (inert gas) or induce internal pressure in the molded part to help hold it out against the mold during the cooling process. Gas injection requires a machine to be fitted with special equipment to be able to inject the gas during the biaxial rotation used in rotational molding.

gel   A resin particle of unusually high molecular weight. A stage of cross-linking at which a molten thermosetting (cross-linkable) resin is about to lose the capacity to flow.

gelation   The stage where a plastisol, when heated, goes from the liquid to the solid state.

gelation temperature   Temperature at which plastisol solidifies. Since this temperature is time-dependent, the procedure for determining it must be specified.

glass transition temperature   The approximate midpoint of the temperature range at which a primarily noncrystalline polymer changes from brittle (glass) to rubbery.

gloss   The shine or luster of a surface.

granular   A form of resin having a nonregular particle size between a powder and a pellet.

granulator   A machine used to grind (reduce plastics parts) to a flake form; the size of the regrind (flakes) is determined by the hole size in the granulator screen. A large hole size will produce large regrind and high rates, but will make the regrind more difficult to pulverize.

grinder   A device used to pulverize plastic resin into an acceptable powder for rotational molding.

grit-blasting   A method of roughing up the cavity surface, and also a means of removing flash from a molded part. In the case of the cavity, sand or steel shot is usually used. For flash removal, sand or finely ground nut pits are used.

guide pin   A projection on one flange of a flat-flanged mold that matches a recess or hole in the mating flange. Used to assure proper alignment of the two flanges.

H **hardness**   The resistance to surface indentation usually measured by the depth of penetration of a blunt point under a given load using a particular instrument according to a prescribed procedure.

haze   Indefinite cloudy appearance within or on a surface of a plastic, different from chalking or bloom.

HDPE   High-density polyethylene.

head-to-head   Molding of one large part that is separated into two individual parts after molding.

heat deflectors   Fins used to direct hot air into recessed areas of molds that would otherwise not get adequate heat during the oven cycle.

heat distortion temperature   The temperature at which a standard test bar deflects 2.54 mm (0.010 in.) under a stated load of either 0.455 or 1.82 MPa (66 or 264 psi) (ASTM method D-648).

heat pin   A special pin capable of transferring heat rapidly. Used in molds to transfer heat to and from hard to heat and cool areas of a mold.

heat ribs   Ribs or protrusions added to a mold to pick up additional heat in that area of the mold during the molding cycle.

heat sink   A mass of material that draws heat away from nearby parts of a mold, either intentionally (for shielding purposes) or unintentionally, causing thin areas in a molded part.

heat stability   The resistance of polymers, including plastisols, to degradation by heat during fabrication or in use. Evidence of degradation is discoloration to brown and then black.

high-density polyethylene (HDPE)   A polyethylene with a density of 0.941 or higher and usually made by a low-pressure process. HDPE is made by polymerizing ethylene alone, has almost no side groups (branches), and usually has a density of 0.960 or above. Copolymers produced by copolymerizing ethylene with butene, hexene, 4-methyl pentene, octene, etc., have densities of 0.941 to about 0.958. Also called linear polyethylene due to the lack of side groups.

homopolymer   A natural or synthetic high polymer derived from a single monomer.

hoop stress   The force per unit area in the wall of a pipe or tank in the circumferential orientation due to internal hydrostatic pressure.

hot-air convection oven   An oven that performs heat transfer by the convection of heat from hot air directed onto the mold surfaces. Oven types are oil, gas, or electric, either directly or indirectly fired.

hot compounding   A thorough mixing process in which both heat and shear are applied to achieve dispersion of additives, color, or other polymers into a polymer.

hot liquid conduction oven   An oven that performs heat transfer by the conduction of heat from a hot liquid spray directed onto the mold surface. The liquid, normally a eutectic salt, is usually heated and recirculated by a remote heating device.

hydrolysis   Chemical decomposition of a substance involving the addition of water.

hygroscopic   Capable of absorbing and retaining environmental moisture.

I   **IBC**   Intermediate bulk container.

impact resistance   Relative susceptibility of plastics to fracture by shock (e.g., as indicated by the energy expended by a standard pendulum or falling-dart impact machine in breaking a standard specimen in one blow) (ASTM method D-256).

impregnation   The process of filling voids in a material such as cast aluminum.

in-mold graphics   A material in decal form that is applied to the mold surface and, during the process, migrates and penetrates into the molded part.

independent arm   A rotational molding machine with arms for molding that index around the center hub independently of each other.

index   The movement of an arm or arms from one station (location) to another. This is generally in a circular motion as on a carousel or ferris-wheel molding machine.

induction time   The portion of the heating cycle from the beginning of the oven cycle to when powder starts to fuse to the inside mold surface.

industrial finish   A low-cost cavity finish that may contain cutter marks and scratches left by sanding.

infrared radiation oven   An oven that uses infrared radiation to heat molds for processing.

infusible   Incapable of being fused by heat (e.g., a resin that has been cross-linked).

inhibitor   A compound that slows or suppresses a chemical reaction.

initiator   Any substance or a form of energy that starts a chemical reaction; a peroxide that is used to start polymerization or cross-linking of a polyethylene resin; heat or light that may start a polymer degrading.

inorganic pigments   Synthetic pigments not occurring naturally in nature. Small particle size, generally translucent.

insert   A metal (or other material) piece part molded into a final molding (e.g., a metal threaded nut). It also can refer to a part of a mold that is removed before part demolding, normally forming an undercut in the part preventing the part from being pulled out of the mold unless the insert is used.

insert holder   A method used to hold an insert in place during the molding cycle.

intermediate bulk container (IBC)   A container intended for the transportation of liquids or solids. IBCs (by UN-based standards) have a volumetric capacity of not more than three cubic meters (3,000 liters, 793 gallons, or 106 cubic feet) and not less than 0.45 cubic meters (450 liters, 119 gallons, or 15.9 cubic feet), or a maximum net mass of not less than 400 kilograms (880 pounds).

internal cooling   A process used to cool the inside of a part while the mold is in the cooler. This process is used to shorten the cooling cycle.

internal mold blanketing   Method through which inert gas or cooling media can be introduced into the mold during the heating and cooling cycles. The inert gas and cooling media are introduced into the mold through hollow rotating drive shafts, mechanical seals, and a flexible high-temperature hose inserted in the mold.

ISO   International Organization of Standards.

ISO 9000   A series of voluntary standards for a quality system.

Izod impact strength   A test designed to determine the resistance of a material to shock loading. It involves the notching of a specimen, which is then placed in the jaws of a machine and struck with a weighted pendulum (ASTM D-256).

K   **K factor**   The coefficient of thermal conductivity. The amount of heat that passes through a specific unit of material in a given time when the difference in temperature of two faces is one degree.

kerksite   A high thermal conductive alloy of aluminum and zinc used for cast molds.

kiss-offs   Internal protrusions in a mold that, when the mold is assembled, touch together, creating a tube through a molded part.

kiss-off ribs   Elongated internal protrusions in a mold that cause two opposite walls of a part to weld to each other during the molding process.

L   **label panel**   A special surface designed into a part on which to apply a label. This commonly requires a textured mold to have a smooth surface in that area. The area may be recessed to locate and protect the label.

latent heat of crystallization   The amount of heat released from a material when it passes from molten form to solid form.

latitude   The response of a material to variations in fabrication conditions. A compound of wide latitude can be fabricated over a considerable range of conditions, and conversely for a compound of narrow latitude.

lay-down time   The time in the rotational molding cycle from when the powder starts to adhere onto the mold surface to when no free-flowing powder is left in the mold.

LDPE   Low-density polyethylene.

lift rings   Attachments on a mold frame to attach a hoist.

light resistance   The ability of a plastic material to resist changes in color or other characteristics during exposure to sunlight or ultraviolet light.

linear low/medium density polyethylene (LLDPE/LMDPE)   A polyethylene copolymer resin produced in a low-pressure reactor with densities below 0.941. The backbone chain is free of branching except for those relatively short chains of uniform length that occur as a result of reacting ethylene with another monomer (e.g., propylene, butene, hexene). This structure provides a resin with more desirable processing and performance properties than are available in LDPE or MDPE, which are produced via the high-pressure process.

linear molecule   A long chain molecule with a continuous backbone, in contrast to one with many side chains or branches. (See high-density polyethylene.)

LLDPE/LMDPE   Linear low/medium density polyethylene.

load/unload station   An index position where loading of the mold with raw material and unloading of molded parts from mold cavities take place.

long chain branching   The occurrence in a polymer of branches that are up to several hundred carbon atoms in length. Such branches occur especially in some low-density polyethylene resins made by the high-pressure process.

low-density polyethylene (LDPE)   Generally refers to polyethylene copolymer produced by the high-pressure process that has a density of 0.910 to 0.925 and is characterized by significant long chain branching.

low-temperature impact test   A test method that measures the ability of a rotationally molded specimen to withstand a given dart impact without rupturing or cracking at a given low temperature. The value is expressed in Joules or in foot pounds of energy.

M major axis   The major axis of a molding machine arm that undergoes biaxial orthogonal motion is the axis in the horizontal direction. Mold motion about this axis will generally be in a vertical plane.

manual mode   A machine mode for a continuous rotational molding machine where all machine functions are controlled through the machine podium by the operator.

masterbatch   A plastics ingredient that includes a high concentration of an additive or additives. Masterbatches are designed for use in appropriate quantities with the base resin or mix so that the correct end concentration is achieved.

material handling system A method of automation to move plastic materials around the factory. An example is a vacuum pump moving resin from a silo to a mixing area or to a rotomolder. Material handling may also include automatic weighing and dosing of resin into the molds.

material reservoir An excess cavity in a mold to permit enough material to be placed in the mold in the bulk form to complete the molding to the wall thickness desired. Sometimes used to make loading a narrow mold easier.

maximum swing The diameter of the largest imaginary sphere, with its center located at the intersection of the minor and major axes, that could be placed in both the oven and cooling chamber, individually.

maximum weight per arm (spindle) The maximum total weight that can be safely supported and set in motion on a single machine arm. This weight includes mold spider or mold frames, molds, counterweights, and material.

MDPE Medium-density polyethylene.

medium-density polyethylene (MDPE) A polyethylene resin with a density greater than 0.926 and not exceeding 0.94.

melt compounding Mixing of plastics with other plastics, fillers, UV absorbers, antioxidants, or color pigments by shear, usually using an extruder. Achieves better dispersion than dry blending.

melt-flow rate The amount of a thermoplastic polymer that will flow through an orifice, usually 3.43 mm (0.135 in.) long by 2.10 mm (0.0825 in.) in diameter, under prescribed conditions of temperature and pressure. The result is expressed in grams per ten minutes. The test result should note the condition of the test, for example, MFR 230 °C/2.16 kg. Usually polyethylene is tested at 190 °C (374 °F) with a 2160 g (0.982 lb.) weight using a 2.10 mm (0.0825 in.) diameter orifice for ten minutes. Polypropylene is usually tested at 230 °C (446 °F) with a 2160 g (0.982 lb.) weight.

melt index The amount, in grams, of a thermoplastic resin that can be forced through an extrusion plastometer with set orifice, weight, temperature, and time. (Various plastics use different temperature/ pressure settings based on ASTM standards.)

melt strength The strength of the plastic while in the molten state. Usually a qualitative term for the ability of the molten plastic to be extended or drawn without tearing or breaking.

melting point The temperature at which solid and liquid forms of a substance are in equilibrium. In common usage, the melting point is taken as the temperature at which the liquid first forms in a small sample as its temperature is increased gradually.

memory   The tendency of a thermoplastic part to change dimensions when its temperature is raised in response to stresses introduced during fabrication.

mesh size   A method to determine resin particle size (ASTM D-1921).

metallocene   A type of catalyst used to make polyolefins.

metallocene polyethylene (mPE)   Polyethylene, usually copolymers, produced with a metallocene catalyst. These polymers are characterized by very regular copolymer molecule distribution and narrow molecular weight. Some mPE polymers have very low density and are elastomeric in nature.

micropellet   Often an alternative to molding thirty-five mesh powder. These very small pellets (0.50 to 0.75 mm) (0.020 to 0.030 in.) are extruded pellets versus the pulverized powder. Micropellets are generally dust-free and it is easier to automate their handling, compared to thirty-five mesh powder.

migration   Movement of a substance from one material to another with which it is in intimate contact.

minor axis   The minor axis of rotation is perpendicular to the major axis of rotation. Mold motion about this axis will generally be in the horizontal plane. See also major axis.

modified   Containing ingredients such as fillers, pigments, or other additives that help to vary the physical properties of a plastic material.

modulus of elasticity   The ratio of stress to strain in an elastically deformed material.

moisture absorption   The attraction and incorporation of water vapor from air by a material.

moisture content (hygroscopic materials)   A measure of the absorbed water content for a given material. Measured value of 0.2% is considered the maximum for good processing.

mold (noun)   A tool in which material is placed and then acted upon by pressure or heat to form a part conforming to the shape of the mold. In the case of rotational molding, material is placed in the mold, heated, and rotated, so the material collects on the inside surface of the mold. Once the total material has collected and cured on the mold, the mold is cooled until the material has set up. The mold is then opened and the part removed. Rotational molds are usually produced by casting (using aluminum), electroforming, machining out of solid material, or fabricating out of sheet stock. The term normally includes all the components (flanges, clamps, mounting plates, etc.) required to run the mold on a machine. (verb) The act of forming parts by using a mold.

mold opening/closing devices Mechanical, pneumatic, or hydraulic devices for opening and closing the molds or mold spiders. Typical devices are chain hoists, toggle clamps, spring clamps, cams, and jack screws.

mold pattern Material used to establish the configuration of a mold. Used over the part pattern to establish the wall thickness of a cast-aluminum mold.

mold preheat The practice of heating specified areas of a mold before it is cycled into the oven. This method is used to help balance the wall thickness on hard-to-heat mold areas.

mold release A lubricant such as wax, powder, or spray, used to coat the cavities of a mold to prevent the part from sticking or hanging up in the mold, and allowing for easy part removal.

mold rotation The rotating of a mold in two directions (perpendicular axes of rotation) simultaneously while being heated. A mold's rotation ratio of major to minor axis is important in obtaining a uniform deposit of powdered resin on the interior surface of the mold.

mold seam The line where parts of the mold make contact when the mold is assembled.

mold shrinkage The difference in dimensions between a mold and a molded part. The measurement should be made when the molded part has had time to complete its shrinking, and with the mold and the molded part at the same room temperature.

molding cycle The time required to produce a plastic part in a molding operation. Measured from a point of one operation to the same point of the first repeat of the operation.

molding powder Plastic molding compound in a powder form (as opposed to granular form).

molecular weight The sum of the atomic weights of all atoms in a molecule.

molecular weight distribution The ratio of the weight average molecular weight to the number average molecular weight gives an indication of the distribution.

monomer A relatively simple compound, usually containing carbon, and of low molecular weight, which can react to form a polymer by combination with itself or with other similar molecules or compounds.

mounting plate The plate attached to the mold spider or frame that permits the mold to be attached to the molding machine.

mPE Metallocene polyethylene.

multiple-cavity mold   A mold that will make more than one piece per part cycle. See also head-to-head.

multiple molds   Several different molds, contrasted with a multiple-cavity mold.

multiple-piece mold   A mold with more than two pieces that are required to be assembled to make a part.

N   **National Sanitation Foundation (NSF)**   US agency that approves plastic pipe, tanks, containers, etc., for use with consumable foods or liquids when they meet certain criteria.

nonrigid plastic   A plastic with a stiffness or apparent modulus of elasticity of not over 345 MPa (50,000 psi) at 25 °C (77 °F) when determined according to ASTM D-747.

notch sensitivity   The extent to which the sensitivity of a material to fracture is increased by the presence of a surface inhomogeneity such as a notch, a sudden change in section, a crack, or a scratch. Low notch sensitivity is usually associated with ductile materials, and high notch sensitivity with brittle materials.

notch sensitivity bent-strip test   A test of the ability of a rotationally molded specimen to withstand bending away from a notched surface without breaking at the notch. This is an industry pragmatic test, and is pass or fail.

notched-Izod   Method to determine impact strength of a plastic after it has a notch in it. A weighted pendulum is used to strike the notch. Some plastics are very sensitive to scratches and cuts and therefore this is a useful test.

NSF   National Sanitation Foundation.

nucleation   The process of crystal formation as a resin cools from the melt.

nylon   Family of thermoplastic resin that consists of recurring amide groups in the main chain. Nylon is characterized as strong, tough, and abrasion-resistant, and has good fatigue strength.

O   **OCT**   Oven cycle time.

offset   Parts of a mold that do not align accurately are said to be offset. Normally refers to parting lines.

offset arm   A machine arm whose mold-mounting plate is offset to one side of the actual spherical mold area.

oil-jacketed molding machine   Machine that supplies hot and cold liquid to biaxially rotating molds. The molds are constructed to have double walls, through which hot and cold liquid heat transfer media are circulated during rotation.

one-step foam   A product containing several components that are molded together to establish a solid outer wall and a foam-filled center. This technology eliminates the use of the drop box.

opaque   An opaque material does not transmit light.

open-celled foam   A cellular plastic in which there is a predominance of interconnected cells.

open (gas) flame heating   Heat applied directly to the mold by a gas-fired burner.

orange peel   An uneven surface resembling that of an orange peel.

organic pigments   Generally derived from naturally occurring minerals. Larger particles, generally opaque.

organosol   A liquid vinyl dispersion in which part of the liquid phase is a volatile solvent.

orientation   The alignment of the molecular structure in polymeric material so as to produce a highly uniform structure.

outgassing   The release of a volatile substance from a molding compound during fabrication or while in use.

oven chamber   An enclosed chamber that supplies the heat to the rotating molds.

oven cycle time (OCT)   The total heating cycle required to heat the mold and fuse the molding resin to a proper cure condition.

P   **paneling**   Distortion of a container occurring during aging or storage, caused by the development of reduced pressure inside the container.

part impact   Impact obtained by dropping a filled or partially filled part from a given height. Dart impact and part impact do not necessarily correlate.

part unloading devices   Mechanical, pneumatic, or hydraulic devices for removing finished parts from the molds. May consist of vacuum cups, mechanical grabbers, etc.

parting line   The separation line between two parts of a mold. Also commonly used to identify the seam or flash line showing on a molded part where the mold has been joined together.

pattern   A replica of a part used to produce an electroformed or cast mold. The pattern may be made to a different scale than the original part to compensate for mold and molding shrinkage.

PC   Polycarbonate.

PE   Polyethylene.

pellets   The form of polymers after extrusion. Pellets range in size from 0.50 to 3.12 mm (0.020 to 0.125 in.). Pellets may be cylindrical, square, or round. Pellets are normally pulverized into thirty-five mesh powder for rotational molding.

percent gel   A method to determine the relative amount of cross-linking in a polyethylene resin. The value is expressed as a percentage of material that is insoluble in refluxing solvent after sixteen hours (ASTM D-2765).

permanence   Resistance of a plastic to appreciable changes in characteristics with time and environment.

permanent release coating   A mold coating used to replace solvent-based mold-release. This coating, usually a fluoropolymer, will assist in part removal.

permeability   The rate of diffusion of a vapor, liquid, or solid through a porous medium without physically or chemically altering it.

pigments   Opaque and translucent organic and inorganic colors that are insoluble in plastics.

pinhole   A very small hole in either a mold or a molded part. May be completely through a part or only partially through the part or mold.

pit   Small, crater-like defect on a mold or part surface.

plaster   A material sometimes used in the process of making cast-aluminum molds. In some cases, more than one mold can be made using the same plaster setup. A shelf life exists that may prevent further use of the plaster setup after approximately six months.

plastic   (adjective) The ability of a material to stretch or deform. (noun) A common polymeric material.

plastic deformation   A change in dimension of an object under load that is not recovered when the load is removed; as opposed to elastic deformation.

plastic memory   A phenomenon of plastic to return to its original molded form. Different plastics possess varying degrees of this characteristic.

plasticize   To impart softness and flexibility in a plastic through incorporation of a plasticizer additive.

plasticizer   An organic compound added to plastic compositions to improve flow and processibility, and to reduce brittleness. Plasticizers lower the glass transition temperature. Increasing plasticizer level will increase softness of the part.

plastisol   A liquid dispersion of vinyl resin in plasticizer. Usually contains ingredients such as stabilizers, pigments, and fillers.

plastometer   Instrument used to run melt-flow index or melt index.

plate-out   A thin coating on a mold surface caused by volatility or incompatibility of various ingredients in a molding compound.

PLC   Programmable logic controller.

pockmarks   Irregularly shaped flat depressions on the outer surface of a rotationally molded part caused by accumulation of gas that lifts the plastic away from the surface. The wall thickness is generally not reduced. This phenomenon is more prevalent in resins that give off decomposition products during molding, such as cross-linked or foamed products. Also known as coining.

podium   An operator's control panel that contains all controls for manual and semiautomatic modes of operation of the machine.

polybutylene polymer   Polymer made with butene as essentially the sole monomer.

polycarbonate (PC) polymer   A thermoplastic polymer formed by the reaction of phosgene with an aliphatic or aromatic dihydroxyl. These resins are characterized by high impact resistance, outstanding resistance to temperature, and transparency.

polyethylene (PE) polymer   A thermoplastic polymer formed by the polymerization of ethylene with or without comonomers present. It is normally a translucent, waxy solid; is tough even at low temperatures; and is generally little affected by water and a wide range of chemicals. It is the most widely used polymer in the rotational molding industry.

polymer   A compound formed by the reaction of simple molecules having functional groups that permit them to combine, proceeding to high molecular weights under suitable conditions.

polymerization   A chemical reaction in which the molecules of monomers are linked together producing a higher molecular weight material, a polymer.

polypropylene (PP) polymer   A thermoplastic olefin polymerized from propylene gas in the presence of a catalyst. The material is a lightweight, tough, and chemically resistant plastic material.

polyvinyl chloride (PVC)   Thermoplastic compounds formed by polymerization or copolymerization of vinyl or vinylidene halides and vinyl esters.

porosity   Internal voids in either a mold or a piece part. Porosity in a mold frequently will result in blow holes in piece parts made from the mold. Porosity commonly shows up in the base material of cast aluminum molds and in welded joints of fabricated molds.

posts   Extensions cast on the outside surface of a cast aluminum mold to which the frame and clamping mechanism are attached so the clamping pressure will be spread evenly over the parting line of the assembled mold. Posts are also found on some fabricated molds for the same purpose.

pourability   The ability of a given amount of powdered resin to pour freely through a given size funnel in a given time. The value is expressed in seconds. Indicates the ability of a material to flow in a mold (ASTM method D-1895). Typical values for free flowing powders are thirty seconds or less.

powder   The form of resin most often used in rotational molding. The industry standard is thirty-five mesh, which means the powder does not exceed thirty-five mesh in size. The actual grind has a particle size distribution of thirty-five mesh to smaller, extremely fine 200 mesh.

powder molding   A general term encompassing rotational molding, slush molding, and centrifugal molding of dry, sinterable powders such as polyethylene, nylon, and polyvinyl chloride. The powders are charged into molds that are heated and manipulated according to the process being used, causing the powders to sinter or fuse into a uniform layer against the internal mold wall.

powder thermal diffusivity   The thermal conductivity divided by the product of the density and the specific heat.

PP   Polypropylene.

precolor   Colorants predispersed in the polymer with an extruder or Banbury mixer, then pelletized and ground into colored powder. See also hot compounding.

predrying   The drying of a resin or molding compound before its introduction into a mold. Some plastic compounds are hygroscopic and require this treatment, particularly after storage in a humid atmosphere.

primary plasticizer   A plasticizer with sufficient affinity for the polymer or resin so that it is considered compatible and therefore may be used as the sole plasticizer.

programmable logic controller (PLC)   An electronic device that can be programmed to control the operation of a piece of manufacturing equipment.

prototype mold   A preproduction mold, usually made as inexpensively as possible, to prove a piece-part concept.

pry-bar pad   Extensions built on a mold to permit the use of a pry bar to help separate the mold halves if the part sticks in the mold. Pry bars are intended to help prevent the use of screwdrivers and other tools from being inserted into the parting line and damaging it.

pulverizer A device used to grind plastic resin into powder with properties acceptable for rotational molding.

PVC Polyvinyl chloride.

Q quenching (thermoplastics) A process of shock-cooling thermoplastic materials from the molten state.

R radius The rounded corners in molds or parts are commonly defined by the radius of the curved section.

regrind Scrap, trim, and rejected pieces of thermoplastic that are ground and fed again to the fabricating equipment. The term regrind suggests that some of the resin has undergone a prior heat history and that physical properties may be reduced.

reinforced plastic mold A prototype of mold commonly made out of epoxy materials reinforced with glass or carbon fibers. These molds are not normally used for prolonged production runs.

reinforced plastics A combination of resin and fibrous material, generally glass. These materials are characterized by their high rigidity.

release agent A lubricant used to coat a mold cavity to prevent the molded piece from sticking to it, and thus to facilitate its removal from the mold.

residual stresses A system of processing induced residual stresses that influence properties such as mechanical, physical, chemical, and aesthetic factors.

resilience The tendency of a material to return to its initial shape after a deforming stress is removed.

resin A solid or semisolid organic product of natural or synthetic origin, generally of high molecular weight, with no definite melting point. Most resins are polymers (but all polymers and plastics are resins). Thermoset resins and thermoplastic polymers are the proper term.

resin-loading device Electromechanical or pneumatic devices for dispensing the raw materials into the open mold. May consist of net weighers, volumetric feeders, surge hoppers, dispensing pumps, etc.

reverse index Turret index in the opposite direction of its normal movement.

reverse rotation Mechanism that by set time sequences, reverses the direction of rotation about the major or minor axes during processing.

rheology The study of flow and the characteristics of flowing materials under stress or specified conditions of time, heat, etc.

rib A configuration included in the design of a mold or part to give it strength or enhance its appearance.

rigid plastics For purposes of general clarification, a plastic that has modulus of elasticity either in flexure or in tension greater than 690 MPa (100,000 psi) at 23 °C (73 °F) and 50% relative humidity when tested in accordance with ASTM methods D-747 or D-790 for stiffness of plastics.

rock-and-roll rotational molding machine A machine that uses a combination of uniaxial mold rotation in one direction, and a rocking motion in the other, while heat is applied to the mold.

Rockwell hardness A test for the harder plastics in which the hardness number is derived from the dent, or permanent deformation, caused by a steel ball. The Rockwell hardness may correlate with resistance to surface marring in end use, but not usually with abrasion or wear resistance (ASTM D-785).

rotational mold A mold specifically designed to use the rotational molding process to make parts.

rotational molding A process used to mold hollow parts. The material is placed in the cavity of a mold that rotates in two axes. The mold is subjected to heating and then cooling while rotating. The material melts and adheres to the cavity walls to form the shape desired. Also called rotational casting.

rotational molding machine A machine that provides biaxial rotation to a mold containing powder or liquid plastic, while taking it through a heating and cooling cycle.

rotational ratio or ratio of rotation This is the ratio of major axis rpm to minor axis rpm, and more specifically the ratio R when both numerator and denominator are reduced to the lowest possible whole numbers by multiplying both by the same quantity.

rotolining A process of plastic coating of the inside surface of a hollow article, mostly metal, by the biaxial rotation of rotational molding.

Rotolog A temperature-measuring device used to measure temperatures of the mold, oven air, polymer, or internal mold air temperature during the heating and cooling cycles as they are in progress.

rotomolding A contraction of the term rotational molding.

S **scission** The breaking of polymer chains, lowering the molecular weight, usually by heat or radiation.

screening (heat blanketing) Selective covering of specific positions on a mold's exterior by screening, sheeting, or other insulating materials. The insulation reduces or eliminates the rate of heat transfer to that area of the mold.

secondary material A material that is used in place of a primary raw material in manufacturing a product.

secondary plasticizer or extender A plasticizer with insufficient affinity for the resin to be compatible as the sole plasticizer. Must be blended with a primary plasticizer. Secondary plasticizers are generally used to reduce the compound cost.

semiautomatic mode A machine-processing mode where a rotational molding machine operates on a continuous basis with the input of the operator through a ready button.

semicrystalline This is the proper term for crystalline in polymers, such as polyethylene and acetal, because no plastics are 100% crystalline.

set To allow the plastics to change from liquid to solid. Interchangeable term with cure.

shear strength The ability of a material to withstand shear stress. The stress at which a material fails in shear.

shelf life The time during which a molding compound, resin, etc., can be stored without losing its suitable physical or chemical properties.

shell mold or cavity A thin-walled mold with an outside shape that follows the shape of the cavity.

shielding A material used on the outside or inside surface of a mold to prevent that part of the mold from receiving as much heat as the rest of the mold. Normally used to reduce material wall thickness in an area of a part that is normally thrown away.

shot blast A type of finish applied to the inside of a mold to provide a textured surface on the outside of a finished piece part.

shot weight The amount of material by weight put in a mold to obtain the desired wall thickness of a finished part. See also charge.

shrink factor The amount of reduction in size of a part from its molten condition in the mold to its room temperature size after shrinkage has been completed. The shrinkage factor is expressed in cm/cm/°C (in./in./°F).

shrinkage The reduction in size of a part from cooling or other factors that affect the part over time.

shuttle rotational molding machine Machine where the spindle and drive mechanisms are mounted on a carriage that moves on a track through the three stations. This generally occurs in a straight line rather than indexing about a single hub.

sink mark   A small depression on a molded part caused by an increase in material shrinkage in that area.

SI system   Abbreviation for the international system of units (metric).

side group   A group of atoms attached to the main backbone of a polymer or compound.

sinter-melt   The stage at which a rotomolding powdered material is a porous, three-dimensional network, and all the powder has attained a temperature equal to or greater than the rotomolding materials' melt temperature.

sintering   In forming articles from fusible powders, the process of holding the powder at a temperature just below its melting point for a period of time. The particles are fused (sintered), but the mass, as a whole, does not melt.

slip agent   Provides surface lubrication during and immediately following processing a plastic. Compounded into the plastic, the additive acts as an internal lubricant that gradually migrates to the surface.

slush casting   A method of forming hollow objects. Widely used for doll parts and squeeze toys, in which a fluid plastic mixture, usually vinyl plastisol, is poured into a hollow mold provided with an opening until the mold is full. Heat, applied to the mold before or after filling, causes a layer of material to gel against the inner mold wall. After the layer has reached the desired thickness, the excess fluid material is poured out, and additional heat is applied to fuse the layer. After cooling, the article is stripped from the mold. Molds for slush casting are thin-walled for rapid heat transfer. Electroformed copper molds or aluminum castings are used most often.

slush molding   A process similar to slush casting but using dry, sinterable powders.

solvation   (PVC) The process of swelling, gelling, or solution of a resin by a solvent of plasticizer.

solvent   A substance, usually a liquid, that dissolves other substances.

solvent resistance   Ability of a plastic to resist swelling and dissolving in a solvent.

SPC   Statistical process control.

specific gravity   The density (mass per unit volume) of any material divided by that of water at a standard temperature, usually 40 °C (104 °F). Since water's density is nearly 1 g/cc, density in grams/cc and specific gravity are numerically nearly equal.

specific heat   The amount of heat required to raise a unit mass of a material one degree under specified conditions.

spew   (PVC) Liquids migrate to the surface leaving an oily residue in some compounds. Spew is increased by humidity, intermediate heats, and stress.

spider   Matching sets of metal framework that have the mating halves of molds affixed to them. Mold halves are clamped together by clamping the spiders together, not the individual molds.

sprayed-metal mold   A mold made by spraying molten metal onto a master until a shell of predetermined thickness is achieved. The shell is then removed and backed up with plaster, cement, casting resin, or other suitable material. This process is primarily used to make sheet-forming molds and occasionally for rotationally molded prototypes.

spring-loaded frame   A framework where the clamping force is placed on springs, which in turn transfer the force to the parting line. The intent is to provide a more even pressure on the parting line.

stabilizer   An ingredient used in formulations with some polymers to assist in maintaining the physical and chemical properties of the compounded materials throughout the processing and service life of the materials (e.g., heat and UV stabilizers).

stack-molding   A method of stacking molds vertically as they are mounted on the machine. Stack-molding normally requires a unique demolding process.

standard laboratory conditions   Generally refers to normal conditions of $23 \pm 2\,°C$ $(73.4 \pm 3.6\,°F)$ and $50 \pm 5\%$ humidity.

station   A position located on an indexing path that is commonly used as a stopping location. These positions are where operations such as heating, cooling, and loading/unloading commonly take place.

statistical process control (SPC)   The use of mathematical statistics to control a manufacturing process.

stir-in resin   A vinyl resin that does not require grinding to effect dispersion in a plastisol or organosol.

straight arm   A machine arm whose mold-mounting plates are located at the center of the actual spherical mold area.

strain   Elastic deformation due to stress. Measured as the change in length per unit of length in a given direction and expressed in percentage, cm/cm, or in./in.

stress   The unit force or component of force at a point in a body acting on a plane through the point. Expressed in MPa or psi.

stress-whitening   Permanent whitening or color change that occurs when a polymer is bent or impacted.

surface enhancer A solvent that is applied to areas of the mold to eliminate pinholes and porosity in finished parts.

surface finish Refers to the internal surface finish of a mold, which provides the outside surface finish of a piece part.

surfactants Chemicals that modify the surface properties of resins to influence the wetting and flow properties of liquids, allowing formation of emulsions of intimate mixtures of normally incompatible substances.

T **tack-off** Noncontinuous interior protrusion in a mold that causes two opposite walls of a part to weld to each other during the molding process.

tear resistance Resistance of a material to a force acting to initiate and then propagate a failure at the edge of a test specimen.

template A device that fits on a molded part to locate holes, routed features, etc. Commonly made by the mold maker as a secondary tool.

temporary mold A mold used for prototype or low-production quantities of a piece part.

tensile strength The pulling strength of a material measured at either the yield point or at break. Values are reported in either instance in units of MPa or psi (ASTM D-638).

texture The surface finish on a part. Normally refers to the depth or configuration of the pattern on a part.

thermal conductivity The rate at which heat will pass through a material. It is related to the specific heat of a material.

thermal decomposition Decomposition resulting from action by heat. It occurs at a temperature at which some components of the material are separating or associating, with a modification of the macro- or microstructure.

thermal degradation Deterioration of a material's physical properties by exposure to heat.

thermal expansion The increase in size of a material due to temperature increase.

thermal shock The change in physical stress that occurs when a material is rapidly changed from hot to cold or cold to hot. A common example is when rotational molds move from the oven at a high temperature into the cooling chamber and are sprayed with cold water.

thermal stress-cracking Crazing and cracking of some thermoplastic resins that results from overexposure to elevated temperatures or thermal shock.

thermocouple A probe of two dissimilar metals that creates a known voltage at an instrument that indicates temperature.

thermoplastic   Capable of being repeatedly softened by heat and hardened by cooling.

thermoset   A resin that will undergo or has undergone a chemical reaction by the action of heat, catalyst, ultraviolet light, etc. Thermosets may not be heated and remolded.

thixotropic   Fluid whose apparent viscosity decreases, with time, to some constant value at any constant rate of shear. If stirring is discontinued, the apparent viscosity increases gradually to the original value.

thread plug   A part of a mold that forms a thread and must be turned out of the mold before the part can be removed.

through hole   A tube through a part of the mold must be designed to provide.

toggle clamp   An over-center clamp used to hold mold parts together.

toll-compounding   The process of sending resin to a compounder to modify the polymer with color or other additives. Toll refers to the fact that the extrusion company does not own the resin and charges by weight to process it.

tongue-and-groove parting line   A parting line in which one half has a recessed area and the second half has a protrusion that fits into the recessed area. This type of parting line provides mold part alignment as well as a parting line function.

tooling feature   A hole or protrusion not functional to the basic piece part, but used later to locate the piece part for secondary work.

total solids   Measure of nonvolatile materials in a compound. Most plastisols are at or near 100% total solids.

translucent   Permitting light to pass through, but not transparent.

transparent   Able to be seen through (e.g., auto tail lights).

transvector   Device used to direct air to the outside of the mold in defined areas. During the heat cycle, it generates additional heat to increase wall thickness in the areas where the heat was directed.

trim   To remove flash or sharp edges and corners mechanically or by hand from a molding.

turret   Located at the central hub on a continuous machine (e.g., carousel or ferris-wheel), it supplies the indexing and rotation motions for the arms.

U **UL**   Underwriters Laboratories.

ultimate strength   The maximum unit stress a material will withstand when subjected to an applied load in a compression, tension, or shear test.

ultraviolet (UV)   Zone of invisible radiation beyond the violet end of the spectrum of visible radiation. Since UV wavelengths are shorter than the visible, their photons have more energy, enough to initiate some chemical reactions and to degrade most plastics.

undercut   A feature of a part that protrudes into or out of the basic part and interferes with the removal of the part from the mold.

unnotched bent-strip test   Used to visually determine if crazing will occur in a rotationally molded specimen when bending 180° away from the inside surface.

UV (ultraviolet) absorber   Additive that absorbs and screens ultraviolet light from the sun, protecting the polymer and therefore extending the plastic's useful outdoor life.

UV (ultraviolet) stabilizer   Any chemical compound that, when admixed with a thermoplastic resin, selectively absorbs UV rays.

V   **vapor blast**   A method used to clean molds. The force of the moving vapor carries minute particles with it and scrubs the surface of a material.

vent   A tube inserted into a mold to relieve pressure in the mold during the heating cycle and to reduce the negative pressure in the mold during the cooling cycle.

Vicat softening point   The temperature at which a flat-ended needle of 1 mm (0.04 in.) circular cross-section will penetrate a thermoplastic specimen to a depth of 1 mm (0.04 in.) under a specified load using a selected uniform rate of temperature rise.

virgin material   A plastic material in the form of pellets, granules, powder, etc., that has not been subjected to heat from processing other than that required for its initial manufacture.

viscoelasticity   Combination of both viscous (liquid) and elastic (solid) properties in a material.

viscosity   The property of resistance to flow exhibited within the body of a material. Relationship between applied shearing stress and resulting rate of strain in shear.

viscosity stability   The ability of a liquid material to maintain its flow properties with age.

viscous flow   A type of fluid movement in which all particles of the fluid flow in a straight line parallel to the axis of a containing pipe or channel, with little or no mixing or turbidity.

void   (mold) A lack of material in the wall of a mold. Commonly caused by volatiles or air trapped in the material during casting. (parts) Gaseous pockets that have been trapped and cured into a laminate or molding; an unfilled space in a cellular plastic substantially larger than the characteristic individual cells.

volatiles   Materials that are capable of being driven off as a vapor during molding.

volume markers   Numbers or other indicators put in a mold to show relative volume levels in the finished part.

W   warpage   Distortion in a plastic part after the molding operation.

water cooling   Method of spraying water directly on a hot mold surface for the purpose of conductive cooling.

water marks   Glossy marks on an otherwise dull surface caused by water in the mold or part interior due to poor vent design, improper cooling cycle, or incorrect water spray nozzles.

weatherometer   Accelerated weather testing device using heat, light, and humidity.

working life   The period during which a compound, after mixing with catalyst, solvent, or other compounding ingredients, remains suitable for its intended use.

X   xenon arc   A commonly used weatherometer used to accelerate outdoor exposure to determine the life of the polymer or its additives.

XLPE   Cross-linked polyethylene.

Y   yield value   The lowest stress at which a material undergoes plastic deformation. Below this stress, the material is elastic; above it, viscous.

Young's modulus   The modulus of elasticity in tension. The ratio of stress to corresponding strain in a material subjected to deformation.

NOTE: Portions of this Glossary of Terms were provided courtesy of the Association of Rotational Molders.

Index